REPRESENTING AND INTERVENING

REPRESENTING
AND INTERVENING

INTRODUCTORY TOPICS IN THE PHILOSOPHY OF
NATURAL SCIENCE

IAN HACKING

CAMBRIDGE
UNIVERSITY PRESS

CAMBRIDGE UNIVERSITY PRESS
Cambridge, New York, Melbourne, Madrid, Cape Town, Singapore, São Paulo, Delhi

Cambridge University Press
32 Avenue of the Americas, New York, NY 10013-2473, USA

www.cambridge.org
Information on this title: www.cambridge.org/9780521238298

First published 1983
20th printing 2007

Printed in the United States of America

A catalog record for this book is available from the British Library.

ISBN 978-0-521-23829-8 hardback
ISBN 978-0-521-28246-8 paperback

For Rachel

'Reality . . . what a concept' – S.V.

Acknowledgements

What follows was written while Nancy Cartwright, of the Stanford University Philosophy Department, was working out the ideas for her book, *How the Laws of Physics Lie*. There are several parallels between her book and mine. Both play down the truthfulness of theories but favour some theoretical entities. She urges that only phenomenological laws of physics get at the truth, while in Part B, below, I emphasize that experimental science has a life more independent of theorizing than is usually allowed. I owe a good deal to her discussion of these topics. We have different anti-theoretical starting points, for she considers models and approximations while I emphasize experiment, but we converge on similar philosophies.

My interest in experiment was engaged in conversation with Francis Everitt of the Hanson Physical Laboratory, Stanford. We jointly wrote a very long paper, 'Which comes first, theory or experiment?' In the course of that collaboration I learned an immense amount from a gifted experimenter with wide historical interests. (Everitt directs the gyro project which will soon test the general theory of relativity by studying a gyroscope in a satellite. He is also the author of *James Clerk Maxwell*, and numerous essays in the *Dictionary of Scientific Biography*.) Debts to Everitt are especially evident in Chapter 9. Sections which are primarily due to Everitt are marked (E). I also thank him for reading the finished text with much deliberation.

Richard Skaer, of Peterhouse, Cambridge, introduced me to microscopes while he was doing research in the Haematological Laboratory, Cambridge University, and hence paved the way to Chapter 11. Melissa Franklin of the Stanford Linear Accelerator taught me about PEGGY II and so provided the core material for Chapter 16. Finally I thank the publisher's reader, Mary Hesse, for many thoughtful suggestions.

Chapter 11 is from *Pacific Philosophical Quarterly* 62 (1981), 305–22. Chapter 16 is adapted from a paper in *Philosophical Topics* 2

(1982). Parts of Chapters 10, 12 and 13 are adapted from *Versuchungen: Aufsätze zur Philosophie Paul Feyerabends* (ed. Peter Duerr), Suhrkamp: Frankfurt, 1981, Bd. 2, pp. 126–58. Chapter 9 draws on my joint paper with Everitt, and Chapter 8 develops my review of Lakatos, *British Journal for the Philosophy of Science* 30 (1979), pp. 381–410. The book began in the middle, which I have called a 'break'. That was a talk with which I was asked to open the April, 1979, Stanford–Berkeley Student Philosophy conference. It still shows signs of having been written in Delphi a couple of weeks earlier.

Contents

Analytical table of contents

explanation. They hold that theories are instruments for predicting phenomena, and for organizing our thoughts. A criticism of 'inference to the best explanation' is developed.

C.S. Peirce said that something is real if a community of inquirers will end up agreeing that it exists. He thought that truth is what scientific method finally settles upon, if only investigation continues long enough. W. James and J. Dewey place less emphasis on the long run, and more on what it feels comfortable to believe and talk about now. Of recent philosophers, H. Putnam goes along with Peirce while R. Rorty favours James and Dewey. These are two different kinds of anti-realism.

T.S. Kuhn and P. Feyerabend once said that competing theories cannot be well compared to see which fits the facts best. This idea strongly reinforces one kind of anti-realism. There are at least three ideas here. Topic-incommensurability: rival theories may only partially overlap, so one cannot well compare their successes overall. Dissociation: after sufficient time and theory change, one world view may be almost unintelligible to a later epoch. Meaning-incommensurability: some ideas about language imply that rival theories are always mutually incomprehensible and never inter-translatable, so that reasonable comparison of theories is in principle impossible.

H. Putnam has an account of the meaning of 'meaning' which avoids meaning-incommensurability. Successes and failures of this idea are illustrated by short histories of the reference of terms such as: glyptodon, electron, acid, caloric, muon, meson.

Putnam's account of meaning started from a kind of realism but has become increasingly pragmatic and anti-realist. These shifts are described and compared to Kant's philosophy. Both Putnam and Kuhn come close to what is best called transcendental nominalism.

thesis is illustrated by a detailed account of a device that produces concentrated beams of polarized electrons, used to demonstrate violations of parity in weak neutral current interactions. Electrons become tools whose reality is taken for granted. It is not thinking about the world but changing it that in the end must make us scientific realists.

Preface

This book is in two parts. You might like to start with the second half, *Intervening*. It is about experiments. They have been neglected for too long by philosophers of science, so writing about them has to be novel. Philosophers usually think about theories. *Representing* is about theories, and hence it is a partial account of work already in the field. The later chapters of Part A may mostly interest philosophers while some of Part B will be more to a scientific taste. Pick and choose: the analytical table of contents tells what is in each chapter. The arrangement of the chapters is deliberate, but you need not begin by reading them in my order.

I call them introductory topics. They are, for me, literally that. They were the topics of my annual introductory course in the philosophy of science at Stanford University. By 'introductory' I do not mean simplified. Introductory topics should be clear enough and serious enough to engage a mind to whom they are new, and also abrasive enough to strike sparks off those who have been thinking about these things for years.

Introduction: rationality

> You ask me, which of the philosophers' traits are idiosyncrasies?
> For example: their lack of historical sense, their hatred of becoming, their Egypticism.
> They think that they show their *respect* for a subject when they dehistoricize it – when they turn it into a mummy.
>
> (F. Nietzsche, *The Twilight of the Idols*, 'Reason in Philosophy', Chapter 1)

Philosophers long made a mummy of science. When they finally unwrapped the cadaver and saw the remnants of an historical process of becoming and discovering, they created for themselves a crisis of rationality. That happened around 1960.

It was a crisis because it upset our old tradition of thinking that scientific knowledge is the crowning achievement of human reason. Sceptics have always challenged the complacent panorama of cumulative and accumulating human knowledge, but now they took ammunition from the details of history. After looking at many of the sordid incidents in past scientific research, some philosophers began to worry whether reason has much of a role in intellectual confrontation. Is it reason that settles which theory is getting at the truth, or what research to pursue? It became less than clear that reason *ought* to determine such decisions. A few people, perhaps those who already held that morality is culture-bound and relative, suggested that 'scientific truth' is a social product with no claim to absolute validity or even relevance.

Ever since this crisis of confidence, rationality has been one of the two issues to obsess philosophers of science. We ask: What do we really know? What should we believe? What is evidence? What are good reasons? Is science as rational as people used to think? Is all this talk of reason only a smokescreen for technocrats? Such questions about ratiocination and belief are traditionally called logic and epistemology. They are *not* what this book is about.

Scientific realism is the other major issue. We ask: What is the world? What kinds of things are in it? What is true of them? What is truth? Are the entities postulated by theoretical physics real, or only

I

constructs of the human mind for organizing our experiments? These are questions about reality. They are metaphysical. In this book I choose them to organize my introductory topics in the philosophy of science.

Disputes about both reason and reality have long polarized philosophers of science. The arguments are up-to-the-minute, for most philosophical debate about natural science now swirls around one or the other or both. But neither is novel. You will find them in Ancient Greece where philosophizing about science began. I've chosen realism, but rationality would have done as well. The two are intertwined. To fix on one is not to exclude the other.

Is either kind of question important? I doubt it. We do want to know what is really real and what is truly rational. Yet you will find that I dismiss most questions about rationality and am a realist on only the most pragmatic of grounds. This attitude does not diminish my respect for the depths of our need for reason and reality, nor the value of either idea as a place from which to start.

I shall be talking about what's real, but before going on, we should try to see how a 'crisis of rationality' arose in recent philosophy of science. This could be 'the history of an error'. It is the story of how slightly off-key inferences were drawn from work of the first rank.

Qualms about reason affect many currents in contemporary life, but so far as concerns the philosophy of science, they began in earnest with a famous sentence published twenty years ago:

History, if viewed as a repository for more than anecdote or chronology, could produce a decisive transformation in the image of science by which we are now possessed.

Decisive transformation – anecdote or chronology – image of science – possessed – those are the opening words of the famous book by Thomas Kuhn, *The Structure of Scientific Revolutions*. The book itself produced a decisive transformation and unintentionally inspired a crisis of rationality.

A divided image

How could history produce a crisis? In part because of the previous image of mummified science. At first it looks as if there was not exactly one image. Let us take a couple of leading philosophers for illustration. Rudolf Carnap and Karl Popper both began their careers in Vienna and fled in the 1930s. Carnap, in Chicago and Los Angeles, and Popper, in London, set the stage for many later debates.

They disagreed about much, but only because they agreed on basics. They thought that the natural sciences are terrific and that physics is the best. It exemplifies human rationality. It would be nice to have a criterion to distinguish such good science from bad nonsense or ill-formed speculation.

Here comes the first disagreement: Carnap thought it is important to make the distinction in terms of language, while Popper thought that the study of meanings is irrelevant to the understanding of science. Carnap said scientific discourse is meaningful; metaphysical talk is not. Meaningful propositions must be *verifiable* in principle, or else they tell nothing about the world. Popper thought that verification was wrong-headed, because powerful scientific theories can never be verified. Their scope is too broad for that. They can, however, be tested, and possibly shown to be false. A proposition is scientific if it is *falsifiable*. In Popper's opinion it is not all that bad to be pre-scientifically metaphysical, for un-falsifiable metaphysics is often the speculative parent of falsifiable science.

The difference here betrays a deeper one. Carnap's verification is from the bottom up: make observations and see how they add up to confirm or verify a more general statement. Popper's falsification is from the top down. First form a theoretical conjecture, and then deduce consequences and test to see if they are true.

Carnap writes in a tradition that has been common since the seventeenth century, a tradition that speaks of the 'inductive sciences'. Originally that meant that the investigator should make precise observations, conduct experiments with care, and honestly record results; then make generalizations and draw analogies and gradually work up to hypotheses and theories, all the time developing new concepts to make sense of and organize the facts. If the theories stand up to subsequent testing, then we know something about the world. We may even be led to the underlying laws of nature. Carnap's philosophy is a twentieth-century version of this attitude. He thought of our observations as the foundations for our knowledge, and he spent his later years trying to invent an inductive logic that would explain how observational evidence could support hypotheses of wide application.

There is an earlier tradition. The old rationalist Plato admired geometry and thought less well of the high quality metallurgy, medicine or astronomy of his day. This respect for deduction became enshrined in Aristotle's teaching that real knowledge – science – is a matter of deriving consequences from first principles by means of demonstrations. Popper properly abhors the idea of first principles but he is often called a deductivist. This is because he thinks there is only one logic – deductive logic. Popper agreed with David Hume, who, in 1739, urged that we have at most a psychological propensity to generalize from experience. That gives no reason or basis for our inductive generalizations, no more than a young man's propensity to disbelieve his father is a reason for trusting the youngster rather than the old man. According to Popper, the rationality of science has nothing to do with how well our evidence 'supports' our hypotheses. Rationality is a matter of method; that method is conjecture and refutation. Form far-reaching guesses about the world, deduce some observable consequences from them. Test to see if these are true. If so, conduct other tests. If not, revise the conjecture or better, invent a new one.

According to Popper, we may say that an hypothesis that has passed many tests is 'corroborated'. But this does not mean that it is well supported by the evidence we have acquired. It means only that this hypothesis has stayed afloat in the choppy seas of critical testing. Carnap, on the other hand, tried to produce a theory of confirmation, analysing the way in which evidence makes hypotheses more probable. Popperians jeer at Carnapians because they have provided no viable theory of confirmation. Carnapians in revenge say that Popper's talk of corroboration is either empty or is a concealed way of discussing confirmation.

Battlefields

Carnap thought that *meanings* and a theory of *language* matter to the philosophy of science. Popper despised them as scholastic. Carnap favoured *verification* to distinguish science from non-science. Popper urged *falsification*. Carnap tried to explicate good reason in terms of a theory of *confirmation*; Popper held that rationality consists in *method*. Carnap thought that knowledge has *foundations*; Popper urged that there are no foundations and that all our knowledge is *fallible*. Carnap believed in *induction*; Popper held that there is no logic except *deduction*.

All this makes it look as if there were no standard 'image' of science in the decade before Kuhn wrote. On the contrary: whenever we find two philosophers who line up exactly opposite on a series of half a dozen points, we know that in fact they agree about almost everything. They share an image of science, an image rejected by Kuhn. If two people genuinely disagreed about great issues, they would not find enough common ground to dispute specifics one by one.

Common ground

Popper and Carnap assume that natural science is our best example of rational thought. Now let us add some more shared beliefs. What they do with these beliefs differs; the point is that they are shared.

Both think there is a pretty sharp distinction between *observation* and *theory*. Both think that the growth of knowledge is by and large *cumulative*. Popper may be on the lookout for refutations, but he thinks of science as evolutionary and as tending towards the one true theory of the universe. Both think that science has a pretty tight *deductive structure*. Both held that scientific terminology is or ought to be rather *precise*. Both believed in the *unity of science*. That means several things. All the sciences should employ the same methods, so that the human sciences have the same methodology as physics. Moreover, at least the natural sciences are part of one science, and we expect that biology reduces to chemistry, as chemistry reduces to physics. Popper came to think that at least part of psychology and the social world did not strictly reduce to the physical world, but Carnap had no such qualms. He was a founder of a series of volumes under the general title, *The Encyclopedia of Unified Science*.

Both agreed that there is a fundamental difference between the *context of justification* and the *context of discovery*. The terms are due to Hans Reichenbach, a third distinguished philosophical emigré of that generation. In the case of a discovery, historians, economists, sociologists, or psychologists will ask a battery of questions: Who made the discovery? When? Was it a lucky guess, an idea filched from a rival, or the pay-off for 20 years of ceaseless toil? Who paid for the research? What religious or social milieu helped or hindered this development? Those are all questions about the context of *discovery*.

Now consider the intellectual end-product: an hypothesis, theory, or belief. Is it reasonable, supported by the evidence, confirmed by experiment, corroborated by stringent testing? These are questions about *justification* or soundness. Philosophers care about justification, logic, reason, soundness, methodology. The historical circumstances of discovery, the psychological quirks, the social interactions, the economic milieux are no professional concern of Popper or Carnap. They use history only for purposes of chronology or anecdotal illustration, just as Kuhn said. Since Popper's account of science is more dynamic and dialectical, it is more congenial to the historicist Kuhn than the flat formalities of Carnap's work on confirmation, but in an essential way, the philosophies of Carnap and Popper are timeless: outside time, outside history.

Blurring an image

Before explaining why Kuhn dissents from his predecessors, we can easily generate a list of contrasts simply by running across the Popper/Carnap common ground and denying everything. Kuhn holds:

There is no sharp distinction between observation and theory.

Science is not cumulative.

A live science does not have a tight deductive structure.

Living scientific concepts are not particularly precise.

Methodological unity of science is false: there are lots of disconnected tools used for various kinds of inquiry.

The sciences themselves are disunified. They are composed of a large number of only loosely overlapping little disciplines many of which in the course of time cannot even comprehend each other. (Ironically Kuhn's best-seller appeared in the moribund series, *The Encyclopedia of Unified Science*.)

The context of justification cannot be separated from the context of discovery.

Science is in time, and is essentially historical.

Is reason in question?

I have so far ignored the first point on which Popper and Carnap agree, namely that natural science is the paragon of rationality, the gemstone of human reason. Did Kuhn think that science is irrational? Not exactly. That is not to say he took it to be 'rational' either. I doubt that he had much interest in the question.

We now must run through some main Kuhnian themes, both to understand the above list of denials, and to see how it all bears on rationality. Do not expect him to be quite as alien to his predecessors as might be suggested. Point-by-point opposition between philosophers indicates underlying agreement on basics, and in some respects Kuhn is point-by-point opposed to Carnap-Popper.

Normal science

Kuhn's most famous word was *paradigm*, of which more anon. First we should think about Kuhn's tidy structure of revolution: *normal science, crisis, revolution, new normal science.*

The normal science thesis says that an established branch of science is mostly engaged in relatively minor tinkering with current theory. Normal science is *puzzle-solving*. Almost any well-worked-out theory about anything will somewhere fail to mesh with facts about the world – 'Every theory is born refuted'. Such failures in an otherwise attractive and useful theory are *anomalies*. One hopes that by rather minor modifications the theory may be mended so as to explain and remove these small counterexamples. Some normal science occupies itself with mathematical articulation of theory, so that the theory becomes more intelligible, its consequences more apparent, and its mesh with natural phenomena more intricate. Much normal science is technological application. Some normal science is the experimental elaboration and clarification of facts implied in the theory. Some normal science is refined measurement of quantities that the theory says are important. Often the aim is simply to get a precise number by ingenious means. This is done neither to test nor confirm the theory. Normal science, sad to say, is not in the confirmation, verification, falsification or conjecture-and-refutation business at all. It does, on the other hand, constructively accumulate a body of knowledge and concepts in some domain.

Crisis and revolution

Sometimes anomalies do not go away. They pile up. A few may come to seem especially pressing. They focus the energies of the livelier members of the research community. Yet the more people work on the failures of the theory, the worse things get. Counter-examples accumulate. An entire theoretical perspective becomes clouded. The discipline is in *crisis*. One possible outcome is an entirely new approach, employing novel concepts. The problematic phenomena are all of a sudden intelligible in the light of these new ideas. Many workers, perhaps most often the younger ones, are converted to the new hypotheses, even though there may be a few hold-outs who may not even understand the radical changes going on in their field. As the new theory makes rapid progress, the older ideas are put aside. A *revolution* has occurred.

The new theory, like any other, is born refuted. A new generation of workers gets down to the anomalies. There is a new normal science. Off we go again, puzzle-solving, making applications, articulating mathematics, elaborating experimental phenomena, measuring.

The new normal science may have interests quite different from the body of knowledge that it displaced. Take the least contentious example, namely measurement. The new normal science may single out different things to measure, and be indifferent to the precise measurements of its predecessor. In the nineteenth century analytical chemists worked hard to determine atomic weights. Every element was measured to at least three places of decimals. Then around 1920 new physics made it clear that naturally occurring elements are mixtures of isotopes. In many practical affairs it is still useful to know that earthly chlorine has atomic weight 35.453. But this is a largely fortuitous fact about our planet. The deep fact is that chlorine has two stable isotopes, 35 and 37. (Those are not the exact numbers, because of a further factor called binding energy.) These isotopes are mixed here on earth in the ratios 75.53% and 24.47%.

'Revolution' is not novel

The thought of a scientific revolution is not Kuhn's. We have long had with us the idea of the Copernican revolution, or of the 'scientific revolution' that transformed intellectual life in the seventeenth century. In the second edition of his *Critique of Pure Reason* (1787), Kant speaks of the 'intellectual revolution' by which Thales or some other ancient transformed empirical mathematics into demonstrative proof. Indeed the idea of revolution in the scientific sphere is almost coeval with that of political revolution. Both became entrenched with the French Revolution (1789) and the revolution in chemistry (1785, say). That was not the beginning, of course. The English had had their 'glorious revolution' (a bloodless one) in 1688 just as it became realized that a scientific revolution was also occurring in the minds of men and women.[1]

Under the guidance of Lavoisier the phlogiston theory of combustion was replaced by the theory of oxidation. Around this time there was, as Kuhn has emphasized, a total transformation in many chemical concepts, such as mixture, compound, element, substance and the like. To understand Kuhn properly we should not fixate on grand revolutions like that. It is better to think of smaller revolutions in chemistry. Lavoisier taught that oxygen is the principle of acidity, that is, that every acid is a compound of oxygen. One of the most powerful of acids (then or now) was called muriatic acid. In 1774 it was shown how to liberate a gas from this. The gas was called dephlogisticated muriatic acid. After 1785 this very gas was inevitably renamed oxygenized muriatic acid. By 1811 Humphry Davy showed this gas is an element, namely chlorine. Muriatic acid is our hydrochloric acid, HCl. It contains no oxygen. The Lavoisier conception of acidity was thereby overthrown. This event was, in its day, quite rightly called a revolution. It even had the Kuhnian feature that there were hold-outs from the old school. The greatest analytical chemist of Europe, J.J. Berzelius (1779–1848), never publicly acknowledged that chlorine was an element, and not a compound of oxygen.

The idea of scientific revolution does not in itself call in question scientific rationality. We have had the idea of revolution for a long time, yet still been good rationalists. But Kuhn invites the idea that every normal science has the seeds of its own destruction. Here is an idea of perpetual revolution. Even that need not be irrational. Could Kuhn's idea of a revolution as switching 'paradigms' be the challenge to rationality?

1 I.B. Cohen, 'The eighteenth century origins of the concept of scientific revolution', *Journal for the History of Ideas* 37 (1976), pp. 257–88.

Paradigm-as-achievement

'Paradigm' has been a vogue word of the past twenty years, all thanks to Kuhn. It is a perfectly good old word, imported directly from Greek into English 500 years ago. It means a pattern, exemplar, or model. The word had a technical usage. When you learn a foreign language by rote you learn for example how to conjugate *amare* (to love) as *amo, amas, amat . . .*, and then conjugate verbs of this class following this model, called the paradigm. A saint, on whom we might pattern our lives, was also called a paradigm. This is the word that Kuhn rescued from obscurity.

It has been said that in *Structure* Kuhn used the word 'paradigm' in 22 different ways. He later focussed on two meanings. One is the paradigm-as-achievement. At the time of a revolution there is usually some exemplary success in solving an old problem in a completely new way, using new concepts. This success serves as a model for the next generation of workers, who try to tackle other problems in the same way. There is an element of rote here, as in the conjugation of Latin verbs ending in *-are*. There is also a more liberal element of modelling, as when one takes one's favourite saint for one's paradigm, or role-model. The paradigm-as-achievement is the role-model of a normal science.

Nothing in the idea of paradigm-as-achievement speaks against scientific rationality – quite the contrary.

Paradigm-as-set-of-shared-values

When Kuhn writes of science he does not usually mean the vast engine of modern science but rather small groups of research workers who carry forward one line of inquiry. He has called this a disciplinary matrix, composed of interacting research groups with common problems and goals. It might number a hundred or so people in the forefront, plus students and assistants. Such a group can often be identified by an ignoramus, or a sociologist, knowing nothing of the science. The know-nothing simply notes who corresponds with whom, who telephones, who is on the preprint lists, who is invited to the innumerable specialist disciplinary gatherings where front-line information is exchanged years before it is published. Shared clumps of citations at the ends of published papers are a good clue. Requests for money are refereed by 'peer reviewers'. Those peers are a rough guide to the disciplinary matrix within one country, but such matrixes are often international.

Within such a group there is a shared set of methods, standards, and basic assumptions. These are passed on to students, inculcated in textbooks, used in deciding what research is supported, what problems matter, what solutions are admissible, who is promoted, who referees papers, who publishes, who perishes. This is a paradigm-as-set-of-shared-values.

The paradigm-as-set-of-shared-values is so intimately linked to paradigm-as-achievement that the single word 'paradigm' remains a natural one to use. One of the shared values is the achievement. The achievement sets a standard of excellence, a model of research, and a class of anomalies about which it is rewarding to puzzle. Here 'rewarding' is ambiguous. It means that within the conceptual constraints set by the original achievement, this kind of work is intellectually rewarding. It also means that this is the kind of work that the discipline rewards with promotion, finance, research students and so forth.

Do we finally scent a whiff of irrationality? Are these values merely social constructs? Are the rites of initiation and passage just the kind studied by social anthropologists in parts of our own and other cultures that make no grand claims to reason? Perhaps, but so what? The pursuit of truth and reason will doubtless be organized according to the same social formulae as other pursuits such as happiness or genocide. The fact that scientists are people, and that scientific societies are societies, does not cast doubt, yet, upon scientific rationality.

Conversion

The threat to rationality comes chiefly from Kuhn's conception of revolutionary shift in paradigms. He compares it to religious conversion, and to the phenomenon of a gestalt-switch. If you draw a perspective figure of a cube on a piece of paper, you can see it as now facing one way, now as facing another way. Wittgenstein used a figure that can be seen now as a rabbit, now as a duck. Religious conversion is said to be a momentous version of a similar phenomenon, bringing with it a radical change in the way in which one feels about life.

Gestalt-switches involve no reasoning. There can be reasoned religious conversion – a fact perhaps more emphasized in a catholic tradition than a protestant one. Kuhn seems to have the 'born-again' view instead. He could also have recalled Pascal, who thought that a good way to become a believer was to live among believers, mindlessly engaging in ritual until it is true.

Such reflections do not show that a non-rational change of belief might not also be a switch from the less reasonable to the more reasonable doctrine. Kuhn is himself inciting us to make a gestalt-switch, to stop looking at development in science as subject solely to the old canons of rationality and logic. Most importantly he suggests a new picture: after a paradigm shift, members of the new disciplinary matrix 'live in a different world' from their predecessors.

Incommensurability

Living in a different world seems to imply an important consequence. We might like to compare the merits of an old paradigm with those of a successor. The revolution was reasonable only if the new theory fits the known facts better than the old one. Kuhn suggests instead that you may not even be able to express the ideas of the old theory in the language of the new one. A new theory is a new language. There is literally no way of finding a theory-neutral language in which to express, and then compare the two.

Complacently, we used to assume that a successor theory would take under its wing the discoveries of its predecessor. In Kuhn's view it may not even be able to express those discoveries. Our old picture of the growth of knowledge was one of accumulation of knowledge, despite the occasional setback. Kuhn says that although any one normal science may be cumulative, science is not in general that way. Typically after a revolution a big chunk of some chemistry or biology or whatever will be forgotten, accessible only to the historian who painfully acquires a discarded world-view. Critics will of course disagree about how 'typical' this is. They will hold – with some justice – that the more typical case is the one where, for example, quantum theory of relativity takes classical relativity under its wing.

Objectivity

Kuhn was taken aback by the way in which his work (and that of others) produced a crisis of rationality. He subsequently wrote that he never intended to deny the customary virtues of scientific theories. Theories should be accurate, that is, by and large fit existing experimental data. They should be both internally consistent and consistent with other accepted theories. They should be broad in scope and rich in consequences. They should be simple in structure, organizing facts in an intelligible way. They should be fruitful, disclosing new events, new techniques, new relationships. Within a normal science, crucial experiments deciding between rival hypotheses using the same concepts may be rare, but they are not impossible.

Such remarks seem a long way from the popularized Kuhn of *Structure*. But he goes on to make two fundamental points. First, his five values and others of the same sort are never sufficient to make a decisive choice among competing theories. Other qualities of judgement come into play, qualities for which there could, in principle, be no formal algorithm. Secondly:

Proponents of different theories are, I have claimed, native speakers of different languages. . . . I simply assert the existence of significant limits to what the proponents of different theories can communicate to each other . . . .Nevertheless, despite the incompleteness of their communication, proponents of different theories can exhibit to each other, not always easily, the concrete technical results available by those who practice within each theory.[2]

When you do buy into a theory, Kuhn continues, you 'begin to speak the language like a native. No process quite like choice has occurred', but you end up speaking the language like a native nonetheless. You don't have two theories in mind and compare them point by point – they are too different for that. You gradually convert, and that shows itself by moving into a new language community.

2 'Objectivity, value judgment, and theory choice', in T.S. Kuhn, *The Essential Tension*, Chicago, 1977, pp. 320–39.

Anarcho-rationalism

Kuhn did not, I think, originally intend to address reason at all. His contemporary Paul Feyerabend is different. His radical ideas often overlap with Kuhn's, but he is a long-time foe of dogmatic rationality. He has called himself an anarchist, but because anarchists often hurt people, he prefers the name Dada-ist. Let there be no canon of rationality, no privileged class of good reasons, and no mind-binding preferred science or paradigm. These moral injunctions flow partly from a conception of human nature. Rationalists systematically try to constrain the free spirit of the human mind. There are many rationalities, many styles of reason, and also many good modes of life where nothing worth calling reason matters very much. On the other hand Feyerabend does not preclude the use of any style of reason and he certainly has his own.

Reactions

Unlike some of Feyerabend's polemics, the main strands of Kuhn's book do not explicitly oppose scientific rationality. They do offer another picture of science. It has been challenged on every point. His histories have been called in question, his generalizations cast in doubt, and his views on language and incommensurability have been fiercely criticized. Some philosophers have taken defensive postures, trying to preserve old ideas. Others attack with a new conception, hoping to better Kuhn. Imre Lakatos is one such. His work is discussed in Chapter 8 below. He thought of himself as revising Popper in the face of Kuhn. He wanted a scientific rationality free of Kuhn's 'mob psychology'. He invented an intriguing 'Methodology of Scientific Research Programmes' not so much to refute Kuhn as to offer an alternative, rationalist, vision of science.

My own attitude to rationality is too much like that of Feyerabend to discuss it further: what follows is about scientific realism, not about rationality. The best brief summary of the current state of rationality play comes from Larry Laudan.

We may conclude from the existing historical evidence that:

(1) Theory transitions are generally non-cumulative, i.e. neither the logical nor empirical content (nor even the confirmed consequences) of earlier theories is wholly preserved when those theories are supplanted by newer ones.

(2) Theories are generally not rejected simply because they have anomalies nor are they generally accepted simply because they are empirically confirmed.

(3) Changes in, and debates about, scientific theories often turn on conceptual issues rather than on questions of empirical support.

(4) The specific and 'local' principles of scientific rationality which scientists utilize in evaluating theories are not permanently fixed, but have altered significantly through the course of science.

(5) There is a broad spectrum of cognitive stances which scientists take towards theories, including accepting, rejecting, pursuing, entertaining, etc. Any theory of rationality which discusses only the first two will be incapable of addressing itself to the vast majority of situations confronting scientists.

(7) Given the notorious difficulties with notions of 'approximate truth' – at both the semantic and epistemic levels – it is implausible that characterizations of scientific progress which view evolution towards greater truth–likeness as the central aim of science will allow one to represent science as a rational activity.

(8) The co-existence of rival theories is the rule rather than the exception, so that theory evaluation is primarily a comparative affair.[3]

Laudan thinks that scientific rationality resides in the power of science to solve problems. Theory T is to be preferred to theory $T^\star$ when T solves more problems than $T^\star$. We ought not to worry whether T is closer to the truth than $T^\star$ (point 7). Theories can be evaluated only by comparing their ability to solve problems (point 8). Mesh with experimental facts is not the only thing that counts, but also ability to resolve conceptual problems (point 3). It may be rational to pursue research based on ideas that do not square with present information, for the research gets its value from its ongoing problem-solving (point 2).

We need not subscribe to all Laudan's points. I share with critics a doubt that we can compare problem-solving ability. For me, Laudan's most important observation is (point 5): accepting and rejecting theories is a rather minor part of science. Hardly anybody ever does that. I draw a conclusion opposite to Laudan's: rationality is of little moment in science. The philosopher of language, Gilbert Ryle, remarked long ago that it is not the word 'rational' that works for us, but rather the word 'irrational'. I never say of my wise aunt Patricia that she is rational (rather she is sensible, wise, imaginative, perceptive). I do say of my foolish uncle Patrick that he is

---

3 L. Laudan, 'A problem solving approach to scientific progress', in I. Hacking (ed.), *Scientific Revolutions*, Oxford, 1981, pp. 144f.

sometimes irrational (as well as being idle, reckless, confused, unreliable). Aristotle taught that humans are rational animals, which meant that they are able to reason. We can assent to that without thinking that 'rational' is an evaluative word. Only 'irrational', in our present language, is evaluative, and it may mean nutty, unsound, vacillating, unsure, lacking self-knowledge, and much else. The 'rationality' studied by philosophers of science holds as little charm for me as it does for Feyerabend. Reality is more fun, not that 'reality' is any better word. Reality . . . what a concept.

Be that as it may, see how historicist we have become. Laudan draws his conclusions 'from the existing historical evidence'. The discourse of the philosophy of science has been transformed since the time that Kuhn wrote. No longer shall we, as Nietzsche put it, show our respect for science by dehistoricizing it.

Rationality and scientific realism

So much for standard introductory topics in the philosophy of science that will *not* be discussed in what follows. But of course reason and rationality are not so separable. When I do take up matters mentioned in this introduction, the emphasis is always on realism. Chapter 5 is about incommensurability, but only because it contains the germs of irrealism. Chapter 8 is about Lakatos, often regarded as a champion of rationality, but he occurs here because I think he is showing one way to be a realist without a correspondence theory of truth.

Other philosophers bring reason and reality closer together. Laudan, for example, is a rationalist who attacks realist theories. This is because many wish to use realism as the basis of a theory of rationality, and Laudan holds that to be a terrible mistake. In the end I come out for a sort of realism, but this is not at odds with Laudan, for I would never use realism as a foundation for 'rationality'.

Conversely Hilary Putnam begins a 1982 book, *Reason, Truth and History*, by urging 'that there is an extremely close connection between the notions of *truth* and *rationality*'. (Truth is one heading under which to discuss scientific realism.) He continues, 'to put it even more crudely, the only criterion for what is a fact is what it is *rational* to accept' (p. x). Whether Putnam is right or wrong,

Nietzsche once again seems vindicated. Philosophy books in English once had titles such as A.J. Ayer's 1936 *Language, Truth and Logic*. In 1982 we have *Reason, Truth and History*.

It is not, however, history that we are now about to engage in. I shall use historical examples to teach lessons, and shall assume that knowledge itself is an historically evolving entity. So much might be part of a history of ideas, or intellectual history. There is a simpler, more old-fashioned concept of history, as history not of what we think but of what we do. That is not the history of ideas but history (without qualification). I separate reason and reality more sharply than do Laudan and Putnam, because I think that reality has more to do with what we do in the world than with what we think about it.

PART A
REPRESENTING

1   What is scientific realism?

Scientific realism says that the entities, states and processes described by correct theories really do exist. Protons, photons, fields of force, and black holes are as real as toe-nails, turbines, eddies in a stream, and volcanoes. The weak interactions of small particle physics are as real as falling in love. Theories about the structure of molecules that carry genetic codes are either true or false, and a genuinely correct theory would be a true one.

Even when our sciences have not yet got things right, the realist holds that we often get close to the truth. We aim at discovering the inner constitution of things and at knowing what inhabits the most distant reaches of the universe. Nor need we be too modest. We have already found out a good deal.

Anti-realism says the opposite: there are no such things as electrons. Certainly there are phenomena of electricity and of inheritance but we construct theories about tiny states, processes and entities only in order to predict and produce events that interest us. The electrons are fictions. Theories about them are tools for thinking. Theories are adequate or useful or warranted or applicable, but no matter how much we admire the speculative and technological triumphs of natural science, we should not regard even its most telling theories as true. Some anti-realists hold back because they believe theories are intellectual tools which cannot be understood as literal statements of how the world is. Others say that theories must be taken literally – there is no other way to understand them. But, such anti-realists contend, however much we may use the theories we do not have compelling reasons to believe they are right. Likewise anti-realists of either stripe will not include theoretical entities among the kinds of things that really exist in the world: turbines yes, but photons no.

We have indeed mastered many events in nature, says the anti-realist. Genetic engineering is becoming as commonplace as manufacturing steel, but do not be deluded. Do not suppose that long chains of molecules are really there to be spliced. Biologists may think more clearly about an amino acid if they build a molecular model out of wire and coloured balls. The model may help us arrange the phenomena in our minds. It may suggest new microtechnology, but it is not a literal picture of how things really are. I could make a model of the economy out of pulleys and levers and ball bearings and weights. Every decrease in weight M (the 'money supply') produces a decrease in angle I (the 'rate of inflation') and an increase in the number N of ball bearings in this pan (the number of unemployed workers). We get the right inputs and outputs, but no one suggests that this is what the economy *is*.

If you can spray them, then they are real

For my part I never thought twice about scientific realism until a friend told me about an ongoing experiment to detect the existence of fractional electric charges. These are called quarks. Now it is not the quarks that made me a realist, but rather electrons. Allow me to tell the story. It ought not to be a simple story, but a realistic one, one that connects with day to day scientific research. Let us start with an old experiment on electrons.

The fundamental unit of electric charge was long thought to be the electron. In 1908 J.A. Millikan devised a beautiful experiment to measure this quantity. A tiny negatively charged oil droplet is suspended between electrically charged plates. First it is allowed to fall with the electric field switched off. Then the field is applied to hasten the rate of fall. The two observed terminal velocities of the droplet are combined with the coefficient of viscosity of the air and the densities of air and oil. These, together with the known value of gravity, and of the electric field, enable one to compute the charge on the drop. In repeated experiments the charges on these drops are small integral multiples of a definite quantity. This is taken to be the minimum charge, that is, the charge on the electrons. Like all experiments, this one makes assumptions that are only roughly correct: that the drops are spherical, for instance. Millikan at first ignored the fact that the drops are not large compared to the mean free path of air molecules so they get bumped about a bit. But the idea of the experiment is definitive.

The electron was long held to be the unit of charge. We use e as the name of that charge. Small particle physics, however, increasingly suggests an entity, called a quark, that has a charge of $1/3\ e$. Nothing in theory suggests that quarks have independent existence; if they do come into being, theory implies, then they react immediately and are gobbled up at once. This has not deterred an ingenious experiment started by LaRue, Fairbank and Hebard at Stanford. They are hunting for 'free' quarks using Millikan's basic idea.

Since quarks may be rare or short-lived, it helps to have a big ball rather than a tiny drop, for then there is a better chance of having a quark stuck to it. The drop used, although weighing less than 10^{-4} grams, is 10^7 times bigger than Millikan's drops. If it were made of oil it would fall like a stone, almost. Instead it is made of a substance called niobium, which is cooled below its superconducting transition temperature of $9°K$. Once an electric charge is set going round this very cold ball, it stays going, forever. Hence the drop can be kept afloat in a magnetic field, and indeed driven back and forth by varying the field. One can also use a magnetometer to tell exactly where the drop is and how fast it is moving.

The initial charge placed on the ball is gradually changed, and, applying our present technology in a Millikan-like way, one determines whether the passage from positive to negative charge occurs at zero or at $\pm 1/3\ e$. If the latter, there must surely be one loose quark on the ball. In their most recent preprint, Fairbank and his associates report four fractional charges consistent with $+1/3\ e$, four with $-1/3\ e$, and 13 with zero.

Now how does one alter the charge on the niobium ball? 'Well, at that stage,' said my friend, 'we spray it with positrons to increase the charge or with electrons to decrease the charge.' From that day forth I've been a scientific realist. *So far as I'm concerned, if you can spray them then they are real.*

Long-lived fractional charges are a matter of controversy. It is not quarks that convince me of realism. Nor, perhaps, would I have been convinced about electrons in 1908. There were ever so many more things for the sceptic to find out: There was that nagging worry about inter-molecular forces acting on the oil drops. Could that be what Millikan was actually measuring? So that his numbers showed nothing at all about so-called electrons? If so, Millikan goes no way towards showing the reality of electrons. Might there be minimum electric charges, but no electrons? In our quark example we have the same sorts of worry. Marinelli and Morpurgo, in a recent preprint, suggest that Fairbank's people are measuring a new electromagnetic force, not quarks. What convinced me of realism has nothing to do with quarks. It was the fact that by now there are standard emitters with which we can spray positrons and electrons – and that is precisely what we do with them. We understand the effects, we understand the causes, and we use these to find out something else. The same of course goes for all sorts of other tools of the trade, the devices for getting the circuit on the supercooled niobium ball and other almost endless manipulations of the 'theoretical'.

What is the argument about?

The practical person says: consider what you use to do what you do. If you spray electrons then they are real. That is a healthy reaction but unfortunately the issues cannot be so glibly dismissed. Anti-realism may sound daft to the experimentalist, but questions about realism recur again and again in the history of knowledge. In addition to serious verbal difficulties over the meanings of 'true' and 'real', there are substantive questions. Some arise from an intertwining of realism and other philosophies. For example, realism has, historically, been mixed up with materialism, which, in one version, says everything that exists is built up out of tiny material building blocks. Such a materialism will be realistic about atoms, but may then be anti-realistic about 'immaterial' fields of force. The dialectical materialism of some orthodox Marxists gave many modern theoretical entities a very hard time. Lysenko rejected Mendelian genetics partly because he doubted the reality of postulated 'genes'.

Realism also runs counter to some philosophies about causation. Theoretical entities are often supposed to have causal powers: electrons neutralize positive charges on niobium balls. The original nineteenth-century positivists wanted to do science without ever speaking of 'causes', so they tended to reject theoretical entities too. This kind of anti-realism is in full spate today.

Anti-realism also feeds on ideas about knowledge. Sometimes it arises from the doctrine that we can know for real only the subjects of sensory experience. Even fundamental problems of logic get involved; there is an anti-realism that puts in question what it is for theories to be true or false.

Questions from the special sciences have also fuelled controversy. Old-fashioned astronomers did not want to adopt a realist attitude to Copernicus. The idea of a solar system might help calculation, but it does not say how the world really is, for the earth, not the sun, they insisted, is the centre of the universe. Again, should we be realists about quantum mechanics? Should we realistically say that particles do have a definite although unknowable position and momentum? Or at the opposite extreme should we say that the 'collapse of the wave packet' that occurs during microphysical measurement is an interaction with the human mind?

Nor shall we find realist problems only in the specialist natural sciences. The human sciences give even more scope for debate. There can be problems about the libido, the super ego, and the transference of which Freud teaches. Might one use psychoanalysis to understand oneself or another, yet cynically think that nothing answers to the network of terms that occurs in the theory? What should we say of Durkheim's supposition that there are real, though by no means distinctly discernible, social processes that act upon us as inexorably as the laws of gravity, and yet which exist in their own right, over and above the properties of the individuals that constitute society? Could one coherently be a realist about sociology and an anti-realist about physics, or vice versa?

Then there are meta-issues. Perhaps realism is as pretty an example as we could wish for, of the futile triviality of basic philosophical reflections. The questions, which first came to mind in antiquity, are serious enough. There was nothing wrong with asking, once, Are atoms real? But to go on discussing such a question may be only a feeble surrogate for serious thought about the physical world.

That worry is anti-philosophical cynicism. There is also philosophical anti-philosophy. It suggests that the whole family of issues about realism and anti-realism is mickey-mouse, founded upon a prototype that has dogged our civilization, a picture of knowledge 'representing' reality. When the idea of correspondence between thought and the world is cast into its rightful place – namely, the grave – will not, it is asked, realism and anti-realism quickly follow?

Movements, not doctrines

Definitions of 'scientific realism' merely point the way. It is more an attitude than a clearly stated doctrine. It is a way to think about the content of natural science. Art and literature furnish good comparisons, for not only has the word 'realism' picked up a lot of philosophical connotations: it also denotes several artistic movements. During the nineteenth century many painters tried to escape the conventions that bound them to portray ideal, romantic, historical or religious topics on vast and energetic canvases. They chose to paint scenes from everyday life. They refused to 'aestheticize' a scene. They accepted material that was trivial or banal. They refused to idealize it, refused to elevate it: they would not even make their pictures picturesque. Novelists adopted this realist stance, and in consequence we have the great tradition in French literature that passes through Flaubert and which issues in Zola's harrowing descriptions of industrial Europe. To quote an unsympathetic definition of long ago, 'a realist is one who deliberately declines to select his subjects from the beautiful or harmonious, and, more especially, describes ugly things and brings out details of the unsavoury sort'.

Such movements do not lack doctrines. Many issued manifestos. All were imbued with and contributed to the philosophical sensibilities of the day. In literature some latterday realism was called positivism. But we speak of movements rather than doctrine, of creative work sharing a family of motivations, and in part defining itself in opposition to other ways of thinking. Scientific realism and anti-realism are like that: they too are movements. We can enter their discussions armed with a pair of one-paragraph definitions, but once inside we shall encounter any number of competing and divergent opinions that comprise the philosophy of science in its present excited state.

Truth and real existence

With misleading brevity I shall use the term 'theoretical entity' as a portmanteau word for all that ragbag of stuff postulated by theories but which we cannot observe. That means, among other things, particles, fields, processes, structures, states and the like. There are two kinds of scientific realism, one for theories, and one for entities.

The question about theories is whether they are true, or are true-or-false, or are candidates for truth, or aim at the truth.

The question about entities is whether they exist.

A majority of recent philosophers worries most about theories and truth. It might seem that if you believe a theory is true, then you automatically believe that the entities of the theory exist. For what is it to think that a theory about quarks is true, and yet deny that there are any quarks? Long ago Bertrand Russell showed how to do that. He was not, then, troubled by the truth of theories, but was worried about unobservable entities. He thought we should use logic to rewrite the theory so that the supposed entities turn out to be logical constructions. The term 'quark' would not denote quarks, but would be shorthand, via logic, for a complex expression which makes reference only to observed phenomena. Russell was then a realist about theories but an anti-realist about entities.

It is also possible to be a realist about entities but an anti-realist about theories. Many Fathers of the Church exemplify this. They believed that God exists, but they believed that it was in principle impossible to form any true positive intelligible theory about God. One could at best run off a list of what God is not – not finite, not limited, and so forth. The scientific-entities version of this says we have good reason to suppose that electrons exist, although no full-fledged description of electrons has any likelihood of being true. Our theories are constantly revised; for different purposes we use different and incompatible models of electrons which one does not think are literally true, but there are electrons, nonetheless.

Two realisms

Realism about entities says that a good many theoretical entities really do exist. Anti-realism denies that, and says that they are fictions, logical constructions, or parts of an intellectual instrument for reasoning about the world. Or, less dogmatically, it may say that we have not and cannot have any reason to suppose they are not fictions. They may exist, but we need not assume that in order to understand the world.

Realism about theories says that scientific theories are either true or false independent of what we know: science at least aims at the truth, and the truth is how the world is. Anti-realism says that theories are at best warranted, adequate, good to work on, acceptable but incredible, or what-not.

Subdivisions

I have just run together claims about reality and claims about what we know. My realism about entities implies both that a satisfactory theoretical entity would be one that existed (and was not merely a handy intellectual tool). That is a claim about entities and reality. It also implies that we actually know, or have good reason to believe in, at least some such entities in present science. That is a claim about knowledge.

I run knowledge and reality together because the whole issue would be idle if we did not *now* have some entities that some of us think really do exist. If we were talking about some future scientific utopia I would withdraw from the discussion. The two strands that I run together can be readily unscrambled, as in the following scheme of W. Newton-Smith's.[1] He notes three ingredients in scientific realism:

1 An *ontological* ingredient: scientific theories are either true or false, and that which a given theory is, is in virtue of how the world is.

2 A *causal* ingredient: if a theory is true, the theoretical terms of the theory denote theoretical entities which are causally responsible for the observable phenomena.

3 An *epistemological* ingredient: we can have warranted belief in theories or in entities (at least in principle).

Roughly speaking, Newton-Smith's causal and epistemological ingredients add up to my realism about entities. Since there are two ingredients, there can be two kinds of anti-realism. One rejects (1); the other rejects (3).

You might deny the ontological ingredient. You deny that theories are to be taken literally; they are not either true or false; they are intellectual tools for predicting phenomena; they are rules for working out what will happen in particular cases. There are many versions of this. Often an idea of this sort is called *instrumentalism* because it says that theories are only instruments.

Instrumentalism denies (1). You might instead deny (3). An

---

1 W. Newton-Smith, 'The underdetermination of theory by data', *Proceedings of the Aristotelian Society*, Supplementary Volume 52 (1978), p. 72.

example is Bas van Fraassen in his book *The Scientific Image* (1980). He thinks theories are to be taken literally – there is no other way to take them. They are either true or false, and which they are depends on the world – there is no alternative semantics. But we have no warrant or need to believe any theories about the unobservable in order to make sense of science. Thus he denies the epistemological ingredient.

My realism about theories is, then, roughly (1) and (3), but my realism about entities is not exactly (2) and (3). Newton-Smith's causal ingredient says that if a theory is true, then the theoretical terms denote entities that are causally responsible for what we can observe. He implies that belief in such entities depends on belief in a theory in which they are embedded. But one can believe in some entities without believing in any particular theory in which they are embedded. One can even hold that no general deep theory about the entities could possibly be true, for there is no such truth. Nancy Cartwright explains this idea in her book *How the Laws of Physics Lie* (1983). She means the title literally. The laws are deceitful. Only phenomenological laws are possibly true, but we may well know of causally effective theoretical entities all the same.

Naturally all these complicated ideas will have an airing in what follows. Van Fraassen is mentioned in numerous places, especially Chapter 3. Cartwright comes up in Chapter 2 and Chapter 12. The overall drift of this book is away from realism about theories and towards realism about those entities we can use in experimental work. That is, it is a drift away from representing, and towards intervening.

Metaphysics and the special sciences

We should also distinguish realism-in-general from realism-in-particular.

To use an example from Nancy Cartwright, ever since Einstein's work on the photoelectric effect the photon has been an integral part of our understanding of light. Yet there are serious students of optics, such as Willis Lamb and his associates, who challenge the reality of photons, supposing that a deeper theory would show that the photon is chiefly an artifact of our present theories. Lamb is not saying that the extant theory of light is plain false. A more profound theory would preserve most of what is now believed about light, but would show that the effects we associate with photons yield, on analysis, to a different aspect of nature. Such a scientist could well be a realist in general, but an anti-realist about photons in particular.

Such localized anti-realism is a matter for optics, not philosophy. Yet N.R. Hanson noticed a curious characteristic of new departures in the natural sciences. At first an idea is proposed chiefly as a calculating device rather than a literal representation of how the world is. Later generations come to treat the theory and its entities in an increasingly realistic way. (Lamb is a sceptic proceeding in the opposite direction.) Often the first authors are ambivalent about their entities. Thus James Clerk Maxwell, one of the creators of statistical mechanics, was at first loth to say whether a gas really is made up of little bouncy balls producing effects of temperature pressure. He began by regarding this account as a 'mere' model, which happily organizes more and more macroscopic phenomena. He became increasingly realist. Later generations apparently regard kinetic theory as a good sketch of how things really are. It is quite common in science for anti-realism about a particular theory or its entities to give way to realism.

Maxwell's caution about the molecules of a gas was part of a general distrust of atomism. The community of physicists and chemists became fully persuaded of the reality of atoms only in our century. Michael Gardner has well summarized some of the strands that enter into this story.[2] It ends, perhaps, when Brownian motion was fully analysed in terms of molecular trajectories. This feat was important not just because it suggested in detail how molecules were bumping into pollen grains, creating the observable movement. The real achievement was a new way to determine Avogadro's number, using Einstein's analysis of Brownian motion and Jean Perrin's experimental techniques.

That was of course a 'scientific', not a 'philosophical', discovery. Yet realism about atoms and molecules was once the central issue for philosophy of science. Far from being a local problem about one kind of entity, atoms and molecules were the chief candidates for real (or merely fictional) theoretical entities. Many of our present positions on scientific realism were worked out then, in connection

2 M. Gardner, 'Realism and instrumentalism in 19th century atomism', *Philosophy of Science* 46 (1979), pp. 1–34.

with that debate. The very name 'scientific realism' came into use at that time.

Realism-in-general is thus to be distinguished from realism-in-particular, with the proviso that a realism-in-particular can so dominate discussion that it determines the course of realism-in-general. A question of realism-in-particular is to be settled by research and development of a particular science. In the end the sceptic about photons or black holes has to put up or shut up. Realism-in-general reverberates with old metaphysics and recent philosophy of language. It is vastly less contingent on facts of nature than any realism-in-particular. Yet the two are not fully separable and often, in formative stages of our past, have been intimately combined.

Representation and intervention

Science is said to have two aims: theory and experiment. Theories try to say how the world is. Experiment and subsequent technology change the world. We represent and we intervene. We represent in order to intervene, and we intervene in the light of representations. Most of today's debate about scientific realism is couched in terms of theory, representation, and truth. The discussions are illuminating but not decisive. This is partly because they are so infected with intractable metaphysics. I suspect there can be no final argument for or against realism at the level of representation. When we turn from representation to intervention, to spraying niobium balls with positrons, anti-realism has less of a grip. In what follows I start with a somewhat old-fashioned concern with realism about entities. This soon leads to the chief modern studies of truth and representation, of realism and anti-realism about theories. Towards the end I shall come back to intervention, experiment, and entities.

The final arbitrator in philosophy is not how we think but what we do.

2 Building and causing

Does the word 'real' have any use in natural science? Certainly. Some experimental conversations are full of it. Here are two real examples. The cell biologist points to a fibrous network that regularly is found on micrographs of cells prepared in a certain way. It looks like chromatin, namely the stuff in the cell nucleus full of fundamental proteins. It stains like chromatin. But it is not real. It is only an artifact that results from the fixation of nucleic sap by glutaraldehyde. We do get a distinctive reproduction pattern, but it has nothing to do with the cell. It is an artifact of the preparation.[1]

To turn from biology to physics, some critics of quark-hunting don't believe that Fairbank and his colleagues have isolated long-lived fractional charges. The results may be important but the free quarks aren't real. In fact one has discovered something quite different; a hitherto unknown new electromagnetic force.

What does 'real' mean, anyway? The best brief thoughts about the word are those of J.L. Austin, once the most powerful philosophical figure in Oxford, where he died in 1960 at the age of 49. He cared deeply about common speech, and thought we often prance off into airy-fairy philosophical theories without recollecting what we are saying. In Chapter 7 of his lectures, *Sense and Sensibilia*, he writes about reality: 'We must not dismiss as beneath contempt such humble but familiar phrases as "not real cream".' That was his first methodological rule. His second was not to look for 'one single specifiable always-the-same *meaning*'. He is warning us against looking for synonyms, while at the same time urging systematic searches for regularities in the usage of a word.

He makes four chief observations about the word 'real'. Two of these seem to me to be important even though they are expressed somewhat puckishly. The two right remarks are that the word 'real'

---

1 For example, R.J. Skaer and S. Whytock, 'Chromatin-like artifacts from nuclear sap', *Journal of Cell Science* 26 (1977), pp. 301–5.

is substantive-hungry: hungry for nouns. The word is also what Austin, in a genially sexist way, calls a trouser-word.

The word is hungry for nouns because 'that's real' demands a noun to be properly understood: real cream, a real constable, a real Constable.

'Real' is called a trouser-word because of negative uses of the words 'wear the trousers'. Pink cream is pink, the same colour as a pink flamingo. But to call some stuff real cream is not to make the same sort of positive assertion. Real cream is, perhaps, not a non-dairy coffee product. Real leather is hide, not naugehyde, real diamonds are not paste, real ducks are not decoys, and so forth. The force of 'real *S*' derives from the negative 'not (a) real *S*'. Being hungry for nouns and being a trouser-word are connected. To know what wears the trousers we have to know the noun, in order that we can tell what is being denied in a negative usage. Real telephones are, in a certain context, not toys, in another context, not imitations, or not purely decorative. This is not because the word is ambiguous, but because whether or not something is a real *N* depends upon the *N* in question. The word 'real' is regularly doing the same work, but you have to look at the *N* to see what work is being done. The word 'real' is like a migrant farm worker whose work is clear: to pick the present crop. But what is being picked? Where is it being picked? How is it being picked? That depends on the crop, be it lettuce, hops, cherries or grass.

In this view the word 'real' is not ambiguous between 'real chromatin', 'real charge', and 'real cream'. One important reason for urging this grammatical point is to discourage the common idea that there *must* be different kinds of reality, just because the word is used in so many ways. Well, perhaps there are different kinds of reality. I don't know, but let not a hasty grammar force us to conclude there are different kinds of reality. Moreover we now must force the philosopher to make plain what contrast is being made by the word 'real' in some specialized debate. If theoretical entities are, or are not, real entities, what contrast is being made?

Materialism

J.J.C. Smart meets the challenge in his book, *Philosophy and Scientific Realism* (1963). Yes, says Smart, 'real' should mark a contrast. Not all theoretical entities are real. 'Lines of force, unlike electrons, *are* theoretical fictions. I wish to say that this table is composed of electrons, etc., just as this wall is composed of bricks' (p. 36). A swarm of bees is made up of bees, but nothing is made up of lines of force. There is a definite number of bees in a swarm and of electrons in a bottle, but there is no definite number of lines of magnetic force in a given volume; only a convention allows us to count them.

With the physicist Max Born in mind, Smart say that the anti-realist holds that electrons do not occur in the series: 'stars, planets, mountains, houses, tables, grains of wood, microscopic crystals, microbes'. On the contrary, says Smart, crystals *are* made up of molecules, molecules of atoms, and atoms are made up of electrons, among other things. So, infers Smart, the anti-realist is wrong. There are at least some real theoretical entities. On the other hand, the word 'real' marks a significant distinction. In Smart's account, lines of magnetic force are not real.

Michael Faraday, who first taught us about lines of force, did not agree with Smart. At first he thought that lines of force are indeed a mere intellectual tool, a geometrical device without any physical significance. In 1852, when he was over 60, Faraday changed his mind. 'I cannot conceive curved lines of force without the condition of physical existence in that intermediate space.'[2] He had come to realize that it is possible to exert a stress on the lines of force, so they had, in his mind, to have real existence. 'There can be no doubt,' writes his biographer, 'that Faraday was firmly convinced that lines of force were real.' This does not show that Smart is mistaken. It does however remind us that some physical conceptions of reality pass beyond the rather simplistic level of building blocks.

Smart is a *materialist* – he himself now prefers the term physicalist. I do not mean that he insists that electrons are brute matter. By now the older ideas of matter have been replaced by more subtle notions. His thought remains, however, based on the idea that material things like stars and tables are built up out of electrons and so forth. The anti-materialist, Berkeley, objecting to the corpuscles of Robert Boyle and Isaac Newton, was rejecting just such a picture. Indeed Smart sees himself as opposed to phenomenalism, a modern version of Berkeley's immaterialism. It is perhaps

2 All quotations from and remarks about Faraday are from L. Pearce Williams, *Michael Faraday, A biography*, London and New York, 1965.

significant that Faraday was no materialist. He is part of that tradition in physics that downplays matter and emphasizes fields of force and energy. One may even wonder if Smart's materialism is an empirical thesis. Suppose that the model of the physical world, due to Leibniz, to Boscovič, to the young Kant, to Faraday, to nineteenth-century energeticists, is in fact far more successful than atomism. Suppose that the story of building blocks gives out after a while. Would Smart then conclude that the fundamental entities of physics are theoretical fictions?

La Realité Physique, the most recent book by the philosophical quantum theorist, Bernard d'Espagnat, is an argument that we can continue to be scientific realists without being materialists. Hence 'real' must be able to mark other contrasts than the one chosen by Smart. Note also that Smart's distinction does not help us say whether the theoretical entities of social or psychological science are real. Of course one can to some extent proceed in a materialistic way. Thus we find the linguist Noam Chomsky, in his book *Rules and Representations* (1980), urging realism in cognitive psychology. One part of his claim is that structured material found in the brain, and passed down from generation to generation, helps explain language acquisition. But Chomsky is not asserting only that the brain is made up of organized matter. He thinks the structures are responsible for some of the phenomenon of thought. Flesh and blood structures in our heads cause us to think in certain ways. This word 'cause' prompts another version of scientific realism.

Causalism

Smart is a materialist. By analogy say that someone who emphasizes the causal powers of real stuff is a *causalist*. David Hume may have wanted to analyse causality in terms of regular association between cause and effect. But good Humeians know there must be more than mere correlation. Every day we read this sort of thing:

While the American College of Obstetricians and Gynecologists recognizes that an association has been established between toxic-shock-syndrome and menstruation-tampon use, we should not assume that this means there is a definite cause-and-effect relationship until we better understand the mechanism that creates this condition. (Press release, October 7, 1980.)

A few young women employing a new brand ('Everything you've ever wanted in a tampon . . . or napkin') vomit, have diarrhoea and a high fever, some skin rash, and die. It is not just fear of libel suits that makes the College want a better understanding of mechanisms before it speaks of causes. Sometimes an interested party does deny that an association shows anything. For example, on September 19, 1980, a missile containing a nuclear warhead blew up after someone had dropped a pipe wrench down the silo. The warhead did not go off, but soon after the chemical explosion the nearby village of Guy, Arkansas, was covered in reddish-brown fog. Within an hour of the explosion the citizens of Guy had burning lips, shortness of breath, chest pains, and nausea. The symptoms continued for weeks and no one anywhere else in the world had the same problem. Cause and Effect? 'The United States Air Force has contended that no such correlation has been determined.' (Press release, October 11, 1980).

The College of Obstetricians and Gynecologists insists that we cannot talk of causes until we find out how the causes of toxic-shock syndrome actually work. The Air Force, in contrast, is lying through its teeth. It is important to the causalist that such distinctions arise in a natural way. We distinguish ludicrous denials of any correlation, from assertions of correlations. We also distinguish correlations from causes. The philosopher C.D. Broad once made this anti-Humeian point in the following way. We may observe that every day a factory hooter in Manchester blows at noon, and exactly at noon the workers in a factory in Leeds lay down their tools for an hour. There is a perfect regularity, but the hooter in Manchester is not the cause of the lunch break in Leeds.

Nancy Cartwright advocates causalism. In her opinion one makes a very strong claim in calling something a cause. We must understand why a certain type of event regularly produces an effect. Perhaps the clearest proof of such understanding is that we can actually use events of one kind to produce events of another kind. Positrons and electrons are thus to be called real, in her vocabulary, since we can for example spray them, separately, on the niobium droplet and thereby change its charge. It is well understood why this effect follows the spraying. One made the experimental device because one knew it would produce these effects. A vast number of very different causal chains are understood and employed. We are entitled to speak of the reality of electrons not because they are building blocks but because we know that they have quite specific causal powers.

This version of realism makes sense of Faraday. As his biographer put it:

> The magnetic lines of force are visible if and when iron filings are spread around a magnet, and the lines are supposedly denser where the filings are thicker. But no one had assumed that the lines of force are there, in reality, even when the iron filings are removed. Faraday now did: we can cut these lines and get a real effect (for example with the electric motor that Faraday invented) – hence they are real.

The true story of Faraday is a little more complicated. Only long after he had invented the motor did he set out his line of force realism in print. He began by saying 'I am now about to leave the strict line of reasoning for a time, and enter upon a few speculations respecting the physical character of lines of force'. But whatever the precise structure of Faraday's thought, we have a manifest distinction between a tool for calculation and a conception of cause and effect. No materialist who follows Smart will regard lines of force as real. Faraday, tinged with immaterialism, and something of a causalist, made just that step. It was a fundamental move in the history of science. Next came Maxwell's electrodynamics that still envelops us.

Entities not theories

I distinguished *realism about entities* and *realism about theories*. Both causalists and materialists care more for entities than theories. Neither has to imagine that there is a best true theory about electrons. Cartwright goes further; she denies that the laws of physics state the facts. She denies that the models that play such a central role in applied physics are literal representations of how things are. She is an anti-realist about theories and a realist about entities. Smart could, if he chose, take a similar stance. We may have no true theory about how electrons go into the build-up of atoms, then of molecules, then of cells. We will have models and theory sketches. Cartwright emphasizes that in several branches of quantum mechanics the investigator regularly uses a whole battery of models of the same phenomena. No one thinks that one of these is the whole truth, and they may be mutually inconsistent. They are intellectual tools that help us understand phenomena and build bits and pieces of experimental technology. They enable us to intervene in processes and to create new and hitherto unimagined phen-

omena. But what is actually 'making things happen' is not the set of
laws, or true laws. There are no exactly true laws to make anything
happen. It is the electron and its ilk that is producing the effects.
The electrons are real, they produce the effects.

This is a striking reversal of the empiricist tradition going back to
Hume. In that doctrine it is only the regularities that are real.
Cartwright is saying that in nature there are no deep and completely
uniform regularities. The regularities are features of the ways in
which we construct theories in order to think about things. Such a
radical doctrine can only be assessed in the light of her detailed
treatment in *How the Laws of Physics Lie*. One aspect of her
approach is described in Chapter 12 below.

The possibility of such a reversal owes a good deal to Hilary
Putnam. As we shall find in Chapters 6 and 7, he had readily
modified his views. What is important here is that he rejects the
plausible notion that theoretical terms, such as 'electron', get their
sense from within a particular theory. He suggests instead that we
can name kinds of things that the phenomena suggest to an
inquiring and inventive mind. Sometimes we shall be naming
nothing, but often one succeeds in formulating the idea of a kind of
thing that is retained in successive elaborations of theory. More
importantly one begins to be able to do things with the theoretical
entity. Early in the day one may start to measure it; much later, one
may spray with it. We shall have all sorts of incompatible accounts
of it, all of which agree in describing various causal powers which
we are actually able to employ while intervening in nature.
(Putnam's ideas are often run together with ideas about essence and
necessity more attributable to Saul Kripke: I attend only to the
practical and pragmatic part of Putnam's account of naming.)

Beyond physics

Unlike the materialist, the causalist can consider whether the
superego or late capitalism is real. Each case has to stand on its own:
one might conclude that Jung's collective unconscious is not real
while Durkheim's collective consciousness is real. Do we sufficiently understand what these objects or processes do? Can we
intervene and redeploy them? Measurement is not enough. We can
measure IQ and boast that a dozen different techniques give the
same stable array of numbers, but we have not the slightest causal

understanding. In a recent polemic Stephen Jay Gould speaks of the 'fallacy of reification' in the history of IQ: I agree.

Causalism is not unknown in the social sciences. Take Max Weber (1864–1920), one of the founding fathers. He has a famous doctrine of ideal types. He was using the word 'ideal' fully aware of its philosophical history. In his usage it contrasts with 'real'. The ideal is a conception of the human mind, an instrument of thought (and none the worse for that). Just like Cartwright in our own day, he was 'quite opposed to the naturalistic prejudice that the goal of the social sciences must be the reduction of reality to "laws"'. In a cautious observation about Marx, Weber writes,

All specifically Marxian 'laws' and developmental constructs, in so far as they are theoretically sound, are ideal types. The eminent, indeed *heuristic* significance of these ideal-types when they are used for the *assessment* of reality is known to everyone who has ever employed Marxian concepts and hypotheses. Similarly their perniciousness, as soon as they are thought of as empirically valid or real (i.e. truly metaphysical) 'effective forces', 'tendencies', etc., is likewise known to those who have used them.[3]

One can hardly invite more controversy than by citing Marx and Weber in one breath. The point of the illustration is, however, a modest one. We may enumerate the lessons:

1 The materialist, such as Smart, can attach no direct sense to the reality of social science entities.

2 The causalist can.

3 The causalist may in fact reject the reality of any entities yet proposed in theoretical social science; materialist and causalist may be equally sceptical – although no more so than the founding fathers.

4 Weber's doctrine of ideal types displays a causalist attitude to social science laws. He uses it in a negative way. He holds that for example Marx's ideal types are not real precisely because they do not have causal powers.

5 The causalist may distinguish some social science from some physical science on the ground that the latter has found some entities whose causal properties are well understood, while the former has not.

3 'Objectivity in social science and social policy', German original 1904, in Max Weber, *The Methodology of the Social Sciences* (E.A. Shils and H.A. Finch, eds. and trans.), New York, 1949, p. 103.

My chief lesson here is that at least some scientific realism can use the word 'real' very much the same way that Austin claims is standard. The word is not notably ambiguous. It is not particularly deep. It is a substantive-hungry trouser-word. It marks a contrast. What contrast it marks depends upon the noun or noun phrase N that it modifies or is taken to modify. Then it depends upon the way that various candidates for being N may fail to be N. If the philosopher is suggesting a new doctrine, or a new context, then one will have to specify why lines of force, or the id, fail to be real entities. Smart says entities are for building. Cartwright says they are for causing. Both authors will deny, although for different reasons, that various candidates for being real entities are, in fact, real. Both are scientific realists about some entities, but since they are using the word 'real' to effect different contrasts, the contents of their 'realisms' are different. We shall now see that the same thing can happen for anti-realists.

3 Positivism

One anti-realist tradition has been around for a long time. At first sight it does not seem to worry about what the word 'real' means. It says simply: there *are* no electrons, nor any other theoretical entities. In a less dogmatic mood it says we have no good reason to suppose that any such things exist; nor have we any expectation of showing that they do exist. Nothing can be known to be real except what might be observed.

The tradition may include David Hume's *A Treatise of Human Nature* (1739). Its most recent distinguished example is Bas van Fraassen's *The Scientific Image* (1980). We find precursors of Hume even in ancient times, and we shall find the tradition continuing long into the future. I shall call it *positivism*. There is nothing in the name, except that it rings a few bells. The name had not even been invented in Hume's day. Hume is usually classed as an empiricist. Van Fraassen calls himself a constructive empiricist. Certainly each generation of philosophers with a positivist frame of mind gives a new form to the underlying ideas and often chooses a new label. I want only a handy way to refer to those ideas, and none serves me better than 'positivism'.

Six positivist instincts

The key ideas are as follows: (1) An emphasis upon *verification* (or some variant such as *falsification*): Significant propositions are those whose truth or falsehood can be settled in some way. (2) *Pro-observation*: What we can see, feel, touch, and the like, provides the best content or foundation for all the rest of our non-mathematical knowledge. (3) *Anti-cause*: There is no causality in nature, over and above the constancy with which events of one kind are followed by events of another kind. (4) *Downplaying explanations*: Explanations may help organize phenomena, but do not provide any deeper answer to *Why* questions except to say that the phenomena regularly occur in such and such a way. (5) *Anti-theoretical entities*:

Positivists tend to be non-realists, not only because they restrict reality to the observable but also because they are against causes and are dubious about explanations. They won't infer the existence of electrons from their causal effects because they reject causes, holding that there are only constant regularities between phenomena. (6) Positivists sum up items (1) to (5) by being *against metaphysics*. Untestable propositions, unobservable entities, causes, deep explanation – these, says the positivist, are the stuff of metaphysics and must be put behind us.

I shall illustrate versions of these six themes by four epochs: Hume (1739), Comte (1830–42), logical positivism (1920–40) and van Fraassen (1980).

Self-avowed positivists

The name 'positivism' was invented by the French philosopher Auguste Comte. His *Course of Positive Philosophy* was published in thick installments between 1830 and 1842. Later he said that he had chosen the word 'positive' to capture a lot of values that needed emphasis at the time. He had, he tells us, chosen the word 'positive' because of its happy connotations. In the major West European languages 'positive' had overtones of reality, utility, certainty, precision, and other qualities that Comte held in esteem.

Nowadays when philosophers talk of 'the positivists' they usually mean not Comte's school but rather the group of logical positivists who formed a famous philosophy discussion group in Vienna in the 1920s. Moritz Schlick, Rudolf Carnap, and Otto Neurath were among the most famous members. Karl Popper, Kurt Gödel, and Ludwig Wittgenstein also came to some of the meetings. The Vienna Circle had close ties to a group in Berlin of whom Hans Reichenbach was a central figure. During the Nazi regime these workers went to America or England and formed a whole new philosophical tradition there. In addition to the figures that I have already mentioned, we have Herbert Feigl and C.G. Hempel. Also the young Englishman A.J. Ayer went to Vienna in the early 1930s and returned to write his marvellous tract of English logical positivism, *Language, Truth and Logic* (1936). At the same time Willard V.O. Quine made a visit to Vienna which sowed the seeds of his doubt about some logical positivist theses, seeds which blossomed into Quine's famous denials of the analytic–synthetic distinction and the doctrine of the indeterminancy of translation.

Such widespread influence makes it natural to call the logical positivists simply positivists. Who remembers poor old Comte, longwinded, stuffy, and not a success in life? But when I am speaking strictly, I shall use the full label 'logical positivism', keeping 'positivism' for its older sense. Among the distinctive traits of logical positivism, in addition to items (1) to (6), is an emphasis on logic, meaning, and the analysis of language. These interests are foreign to the original positivists. Indeed for the philosophy of science I prefer the old positivism just because it is not obsessed by a theory of meaning.

The usual Oedipal reaction has set in. Despite the impact of logical positivism on English-speaking philosophy, no one today wants to be called a positivist. Even logical positivists came to favour the label of 'logical empiricist.' In Germany and France 'positivism' is, in many circles, a term of opprobrium, denoting an obsession with natural science and a dismissal of alternative routes to understanding in the social sciences. It is often wrongly associated with a conservative or reactionary ideology.

In *The Positivist Dispute in German Sociology*, edited by Theodore Adorno, we see German sociology professors and their philosophical peers – Adorno, Jürgens Habermas and so forth – lining up against Karl Popper, whom they call a positivist. He himself rejects that label because he has always dissociated himself from logical positivism. Popper does not share enough of my features (1) to (6) for me to call him a positivist. He is a realist about theoretical entities, and he holds that science tries to discover explanations and causes. He lacks the positivist obsession with observation and the raw data of sense. Unlike the logical positivists he thought that the theory of meaning is a disaster for the philosophy of science. True, he does define science as the class of testable propositions, but far from decrying metaphysics, he thinks that untestable metaphysical speculation is a first stage in the formation of more testable bold conjectures.

Why then did the anti-positivist sociology professors call Popper a positivist? *Because he believes in the unity of scientific method.* Make hypotheses, deduce consequences, test them: that is Popper's method of conjecture and refutation. He denies that there is any peculiar technique for the social sciences, any *Verstehen* that is different from what is best for natural science. In this he is at one with the logical positivists. But I shall keep 'positivism' for the name of an anti-metaphysical collection of ideas (1) to (6), rather than dogma about the unity of scientific methodology. At the same time I grant that anyone who dreads an enthusiasm for scientific rigour will see little difference between Popper and the members of the Vienna Circle.

Anti-metaphysics

Positivists have been good at slogans. Hume set the tone with the ringing phrases with which he concludes his *An Enquiry Concerning Human Understanding*:

When we run over libraries, persuaded of these principles, what havoc must we make? If we take in our hand any volume; of divine or school metaphysics, for instance; let us ask, *Does it contain any abstract reasoning concerning quantity or number?* No. *Does it contain any experimental reasoning concerning matter of fact and existence?* No. Commit it then to the flames: for it can contain nothing but sophistry and illusion.

In the introduction to his anthology, *Logical Positivism*, A.J. Ayer says that this 'is an excellent statement of the positivists' position. In the case of the logical positivists the epithet "logical" was added because they wished to annex the discoveries of modern logic.' Hume, then, is the beginning of the criterion of verifiability intended to distinguish nonsense (metaphysics) from sensible discourse (chiefly science). Ayer began his *Language, Truth and Logic* with a powerful chapter, called 'The elimination of metaphysics'. The logical positivists, with their passion for language and meanings, combined their scorn for idle metaphysics with a meaning-oriented doctrine called 'the verification principle'. Schlick announced that the meaning of a statement is its method of verification. Roughly speaking, a statement was to be meaningful, or to have 'cognitive meaning', if and only if it was verifiable. Surprisingly, no one was ever able to define verifiability so as to exclude all bad metaphysical conversation and include all good scientific talk.

Anti-metaphysical prejudices and a verification theory of meaning are linked largely by historical accident. Certainly Comte was a great anti-metaphysician with no interest in the study of 'meanings'. Equally in our day van Fraassen is as opposed to metaphysics.

He is of my opinion that, whatever be the interest in the philosophy of language, it has very little value for understanding science. At the start of *The Scientific Image*, he writes: 'My own view is that empiricism is correct, but could not live in the linguistic form the [logical] positivists gave it.' (p. 3)

Comte

Auguste Comte was very much a child of the first half of the nineteenth century. Far from casting empiricism into a linguistic form, he was an historicist: that is, he firmly believed in human progress and in the near-inevitability of historical laws. It is sometimes thought that positivism and historicism are at odds with each other: quite the contrary, they are, for Comte, complementary parts of the same ideas. Certainly historicism and positivism are no more necessarily separated than positivism and the theory of meaning are necessarily connected.

Comte's model was a passionate *Essay on the Development of the Human Mind*, left as a legacy to progressive mankind by the radical aristocrat, Condorcet (1743–94). This document was written just before Condorcet killed himself in the cell from which, the following morning, he was to be taken to the guillotine. Not even the Terror of the French Revolution, 1794, could vanquish faith in progress. Comte inherited from Condorcet a structure of the evolution of the human spirit. It is defined by The Law of Three Stages. First we went through a theological stage, characterized by the search for first causes and the fiction of divinities. Then we went through a somewhat equivocal metaphysical stage, in which we gradually replaced divinities by the theoretical entities of half-completed science. Finally we now progress to the stage of positive science.

Positive science allows propositions to count as true-or-false if and only if there is some way of settling their truth values. Comte's *Course of Positive Philosophy* is a grand epistemological history of the development of the sciences. As more and more styles of scientific reasoning come into being, they thereby constitute more and more domains of positive knowledge. Propositions cannot have 'positivity' – be candidates for truth-or-falsehood – unless there is some style of reasoning which bears on their truth value and can at least in principle determine that truth value. Comte, who invented the very word 'sociology', tried to devise a new methodology, a new style of reasoning, for the study of society and 'moral science'. He was wrong in his own vision of sociology, but correct in his meta-conception of what he was doing: creating a new style of reasoning to bring positivity – truth-or-falsehood – to a new domain of discourse.

Theology and metaphysics, said Comte, were earlier stages in human development, and must be put behind us, like childish things. This is not to say that we must inhabit a world denuded of values. In the latter part of his life Comte founded a Positivist Church that would establish humanistic virtues. This Church is not quite extinct; some buildings still stand, a little tatty, in Paris, and I am told that Brazil still possesses strongholds of the institution. Long ago it did flourish in collaboration with other humanistic societies, in many parts of the world. Thus positivism was not only a philosophy of scientism but a new, humanistic, religion.

Anti-cause

Hume notoriously taught that cause is only constant conjunction. To say that A caused B is not to say that A, from some power or character within itself, brought about B. It is only to say that things of type A are regularly followed by things of type B. The details of Hume's argument are analysed in hundreds of philosophy books. We may, however, miss a good deal if we read Hume out of his historical context.

Hume is in fact not responsible for the widespread philosophical acceptance of a constant-conjunction attitude to causation. Isaac Newton did it, unintentionally. The greatest triumph of the human spirit in Hume's day was held to be the Newtonian theory of gravitation. Newton was so canny about the metaphysics of gravity that scholars will debate to the end of time what he really thought. Immediately before Newton, all progressive scientists thought that the world must be understood in terms of mechanical pushes and pulls. But gravity did not seem 'mechanical', for its was action at a distance. For that very reason, Newton's only peer, Leibniz, quite rejected Newtonian gravitation: it was a reactionary reversion to inexplicable occult powers. A positivist spirit triumphed over Leibniz. We learned to think that the laws of gravity are regularities that describe what happens in the world. Then we decided that all causal laws are mere regularities!

For empirically minded people the post-Newtonian attitude was, then, this: we should not seek for causes in nature, but only regularities. We should not think of laws of nature revealing what must happen in the universe, but only what does happen. The natural scientist tries to find universal statements – theories and laws – which cover all phenomena as special cases. To say that we have found the explanation of an event is only to say that the event can be deduced from a general regularity.

There are many classic statements of this idea. Here is one from Thomas Reid's *Essays on the Active Powers of the Human Mind* of 1788. Reid was the founder of what is often called the Scottish School of Common Sense Philosophy, which was imported to form the main American philosophy until the advent of pragmatism at the end of the nineteenth century.

Natural philosophers, who think accurately, have a precise meaning to the terms they use in the science; and, when they pretend to show the cause of any phenomenon of nature, they mean by the cause, a law of nature of which that phenomenon is a necessary consequence.

The whole object of natural philosophy, as Newton expressly teaches, is reducible to these two heads: first, by just induction from experiment and observation, to discover the laws of nature; and then to apply those laws to the solution of the phenomena of nature. This was all that this great philosopher attempted, and all that he thought attainable. (I. vii. 6.)

Comte tells a similar story in his *Cours de philosophie positive*:

The first characteristic of the positive philosophy is that it regards all phenomena as subjected to invariable natural *laws*. Our business is – seeing how vain is any research into what are called *causes*, whether first or final – to pursue an accurate discovery of these laws, with a view to reducing them to the smallest possible number. By speculating upon causes, we could solve no difficulty about origin and purpose. Our real business is to analyze accurately the circumstances of phenomena, and to connect them by the natural relations of succession and resemblance. The best illustration of this is in the case of the doctrine of gravitation. We say that the general phenomena of the universe are *explained* by it, because it connects under one head the whole immense variety of astronomical facts; exhibiting the constant tendency of atoms towards each other in direct proportion to their masses, and in inverse proportion to the squares of their distances; while the general fact itself is a mere extension of one that is perfectly familiar to us and that we therefore say that we know – the weight of bodies on the surface of the earth. As to what weight and attraction are, these are questions that we regard as insoluble, which are not part of positive philosophy and which we rightly abandon to the imagination of the theologians or the subtlety of the metaphysicians. (Paris, 1830, pp. 14–16.)

Logical positivism was also to accept Hume's constant conjunction account of causes. Laws of Nature, in Mortitz Schlick's maxim, *describe* what happens, but do not *prescribe* it. They are accounts of regularities only. The logical positivist account of explanation was finally summed up in C.G. Hempel's 'deductive-nomological' model of explanation. To explain an event whose occurrence is described by the sentence S is to present some laws of nature (i.e. regularities) L, and some particular facts F and to show that the sentence S is deducible from sentences stating L and F. Van Fraassen, who has an interestingly more sophisticated account of explanation, shares the traditional positivist hostility to causes. 'Flights of fancy' he dismissively calls them in his book (for causes are even worse, in his book, than explanation).

Anti-theoretical-entities

Opposition to unobservable entities goes hand in hand with an opposition to causes. Hume's scorn for the entity-postulating sciences of his day is, as always, stated in an ironic prose. He admires the seventeenth-century chemist Robert Boyle for his experiments and his reasoning, but not for his corpuscular and mechanical philosophy that imagines the world to be made up of little bouncy balls or springlike tops. In Chapter LXII of his great *History of England* he tells us that, 'Boyle was a great partisan of the mechanical philosophy, a theory which, by discovering some of the secrets of nature and allowing us to imagine the rest, is so agreeable to the natural vanity and curiosity of men.' Isaac Newton, 'the greatest and rarest genius that ever arose for the ornament and instruction of the species', is a better master than Boyle: 'While Newton seemed to draw off the veil from some of the mysteries of nature, he showed at the same time the imperfections of the mechanical philosophy, and thereby restored her ultimate secrets to that obscurity in which they ever did and ever will remain.'

Hume seldom denies that the world is run by hidden and secret causes. He denies that they are any of our business. The natural vanity and curiosity of our species may let us seek fundamental particles, but physics will not succeed. Fundamental causes ever did and ever will remain cloaked in obscurity.

Opposition to theoretical entities runs through all positivism. Comte admitted that we cannot merely generalize from observations, but must proceed through hypotheses. These must, however, be regarded only as hypotheses, and the more that they postulate, the further they are from positive science. In practical terms, Comte was opposed to the Newtonian aether, soon to be electromagnetic aether, filling all space. He was equally opposed to the atomic hypothesis. You win one, you lose one.

The logical positivists distrusted theoretical entities in varying degrees. The general strategy was to employ logic and language. They took a leaf from Bertrand Russell's notebook. Russell thought that whenever possible, inferred entities should be replaced by logical constructions. That is, a statement involving an entity whose existence is merely inferred from data is to be replaced by a logically equivalent statement about the data. In general these data are closely connected with observation. Thereby arose a great programme of reductionism for the logical positivists, who hoped that all statements involving theoretical entities would by means of logic be 'reduced' to statements that did not make reference to such entities. The failure of this project was greater even than the failure to state the verification principle.

Van Fraassen continues the positivist antipathy to theoretical entities. Indeed he will not even let us speak of theoretical entities: we mean, he writes, simply unobservable entities. These, not being seen, must be inferred. It is van Fraassen's strategy to block every inference to the truth of our theories or the existence of their entities.

Believing

Hume did not believe in the invisible bouncy balls or atoms of Robert Boyle's mechanical philosophy. Newton had showed us that we ought only to seek natural laws that connect the phenomena. We should not allow our natural vanity to imagine that we can successfully seek out causes.

Comte equally disbelieved in the atoms and aether of the science of his time. We need to make hypotheses in order to tell us where to investigate nature, but positive knowledge must lie at the level of the phenomena whose laws we may determine with precision. This is not to say that Comte was ignorant of science. He was trained by the great French theoretical physicists and applied mathematicians. He believed in their laws of phenomena and distrusted any drive towards postulating new entities.

Logical positivism had no such simplistic opportunities.

Members of the Vienna Circle believed the physics of their day: some had made contributions to it. Atomism and electromagnetism had long been established, relativity was a proven success and the quantum theories were advancing by leaps and bounds. Hence arose, in the extreme version of logical positivism, a doctrine of reductionism. It was proposed that in principle there are logical and linguistic transformations in the sentences of theories that will reduce them to sentences about phenomena. Perhaps when we speak of atoms and currents and electric charges we are not to be understood quite literally, for the sentences we use are reducible to sentences about phenomena. Logicians did to some extent oblige. F.P. Ramsey showed how to leave out the names of theoretical entities in the theories, using instead a system of quantifiers. William Craig proved that for any axiomatizable theory involving both observational and theoretical terms, there exists an axiomatizable theory involving only the observational terms. But these results did not do quite what logical positivism wanted, nor was there any linguistic reduction for any genuine science. This was in terrible contrast to the remarkable partial successes by which more superficial scientific theories have been reduced to deeper ones, for example, the ways in which analytic chemistry is founded upon quantum chemistry, or the theory of the gene has been transformed into molecular biology. Attempts at scientific reduction – reducing one empirical theory to a deeper one – have scored innumerable partial successes, but attempts at linguistic reduction have got nowhere.

Accepting

Hume and Comte took all that stuff about fundamental particles and said: We don't believe it. Logical positivism believed it, but said in a sense that it must not be taken literally; our theories are really talking about phenomena. Neither option is open to a present-day positivist, for the programmes of linguistic reduction failed, while on the other hand one can hardly reject the whole body of modern theoretical science. Yet van Fraassen finds a way through this impasse by distinguishing belief from acceptance.

Against the logical positivists, van Fraassen says that theories are to be taken literally. There is no other way to take them! Against the realist he says that we need not believe theories to be true. He invites us instead to use two further concepts: *acceptance* and *empirical adequacy.* He defines scientific realism as the philosophy that maintains that, 'Science aims to give us, in its theories, a literally true story of what the world is like; and acceptance of a scientific theory involves the belief that it is true' (p. 8). His own *constructive empiricism* asserts instead that, 'Science aims to give us theories which are empirically adequate; and acceptance of a theory involves as belief only that it is empirically adequate' (p. 12).

'There is,' he writes, 'no need to believe good theories to be true, nor to believe *ipso facto* that the entities they postulate are real.' The '*ipso facto*' reminds us that van Fraassen does not much distinguish realism about theories from realism about entities. I say that one could believe entities to be real, not 'in virtue of the fact' that one believes some theory to be true, but for other reasons.

A little later van Fraassen explains as follows: 'to accept a theory is (for us) to believe that it is empirically adequate – that what the theory says *about what is observable* (by us) is true' (p. 18). Theories are intellectual instruments for prediction, control, research and sheer enjoyment. Acceptance means commitment, among other things. To accept a theory in your field of research is to be committed to developing the programme of inquiry that it suggests. You may even accept that it provides explanations. But you must reject what has been called inference to the best explanation: to accept a theory because it makes something plain is not thereby to think that what the theory says is literally true.

Van Fraassen's is the most coherent present-day positivism. It has all six features by which I define positivism, and which are shared by Hume, Comte and the logical positivists. Naturally it lacks Hume's psychology, Comte's historicism, and logical positivism's theories of meaning, for those have nothing essential to do with the positivist spirit. Van Fraassen shares with his predecessors the *anti-metaphysics*: 'The assertion of empirical adequacy is a great deal weaker than the assertion of truth, and the restraint to acceptance delivers us from metaphysics' (p. 69). He is *pro-observation*, and *anti-cause*. He *downplays explanation*; he does not think explanation leads to truth. Indeed, just like Hume and Comte, he cites the classic case of Newton's inability to explain gravity as proof that science is not essentially a matter of explanation (p. 94). Certainly he is *anti-theoretical-entities*. So he holds five of our six positivist doctrines. The only one left is the emphasis on *verification* or some variant. Van Fraassen does not subscribe to the logical positivist verifiability theory of meaning. Nor did Comte. Nor, I think, did Hume, although Hume did have an unverifiability maxim for burning books. The positivist enthusiasm for verifiability was only temporarily connected with meaning, in the days of logical positivism. More generally it represents a desire for positive science, for knowledge that can be settled as true, and whose facts are determined with precision. Van Fraassen's constructive empiricism shares this enthusiasm.

Anti-explanation

Many positivist theses were more attractive in Comte's day than our own. In 1840, theoretical entities were thoroughly hypothetical, and distaste for the merely postulated is the starting point for some sound philosophy. But increasingly we have come even to see what was once merely postulated: microbes, genes, even molecules. We have also learned how to use many theoretical entities in order to manipulate other parts of the world. These grounds for realism about entities are discussed in Chapters 10 and 16 below. However one positivist theme stands up rather well: caution about explanation.

The idea of 'inference to the best explanation' is quite old. C.S. Peirce (1839–1914) called it the method of hypothesis, or abduction. The idea is that if, confronted by some phenomenon, you find one explanation (perhaps with some initial plausibility) that makes sense of what is otherwise inexplicable, then you should conclude that the explanation is probably right. At the start of his career Peirce thought that there are three fundamental modes of scientific inference: deduction, induction and hypothesis. The older he got the more sceptical he became of the third category, and by the end of his life he attached no weight at all to 'inference to the best explanation'.

Was Peirce right to recant so thoroughly? I think so, but we need not decide that now. We are concerned only with inference to the best explanation as an argument for realism. The basic idea was enunciated by H. Helmholtz (1821–94), the great nineteenth-century contributor to physiology, optics, electrodynamics and other sciences. Helmholtz was also a philosopher who called realism

'an admirably useful and precise hypothesis'.[1] By now there appear
to be three distinct arguments in circulation. I shall call them the
simple inference argument, the cosmic accident argument, and the
success of science argument.

I am sceptical of all three. I should begin by saying that
explanation may play a less central a role in scientific reasoning than
some philosophers imagine. Nor is *the* explanation of a phenomenon one of the ingredients of the universe, as if the Author of
Nature had written down various things in the Book of the World –
the entities, the phenomena, the quantities, the qualities, the laws,
the numerical constants, and also the explanations of events.
Explanations are relative to human interests. I do not deny that
explaining – 'feeling the key turn in the lock' as Peirce put it – does
happen in our intellectual life. But that is largely a feature of the
historical or psychological circumstances of a moment. There are
times when we feel a great gain in understanding by the organization of new explanatory hypotheses. But that feeling is not a
ground for supposing that the hypothesis is true. Van Fraassen and
Cartwright urge that being an explanation is never a ground for
belief. I am less stringent than they: it seems to me like Peirce to be
merely a feeble ground. In 1905 Einstein explained the photoelectric effect with a theory of photons. He thereby made attractive
the notion of quantized bundles of light. But the ground for
believing the theory is its predictive success, and so forth, not its
explanatory power. Feeling the key turn in the lock makes you feel
that you have an exciting new idea to work with. It is not a ground
for the truth of the idea: that comes later.

Simple inference

The simple inference argument says it would be an absolute miracle
if for example the photoelectric effect went on working while there
were no photons. The explanation of the persistence of this
phenomenon – the one by which television information is converted
from pictures into electrical impulses to be turned into electromagnetic waves in turn to be picked up on the home receiver – is

---

1 'On the aim and progress of physical science' (German original 1871) in H. von Helmholtz,
  Popular Lectures and Addresses on Scientific Subjects (D. Atkinson trans.), London, 1873,
  p. 247.

that photons do exist. As J.J.C. Smart expresses the idea: 'One would have to suppose that there were innumerable lucky accidents about the behavior mentioned in the observational vocabulary, so that they behaved miraculously *as if* they were brought about by the non-existent things ostensibly talked about in the theoretical vocabulary.'[2] The realist then infers that photons are real because otherwise we could not understand how scenes are turned into electronic messages.

Even if, contrary to what I have said, explanation were a ground for belief, this seems not to be an inference to the best explanation at all. That is because the *reality* of photons is no part of the explanation. There is not, after Einstein, some further explanation, namely 'and photons are real', or 'there exist photons'. I am inclined to echo Kant, and say that existence is a merely logical predicate that adds nothing to the subject. To add 'and photons are real', after Einstein has finished, is to add nothing to the understanding. It is not in any way to increase or enhance the explanation.

If the explainer protests, saying that Einstein himself asserted the existence of photons, then he is begging the question. For the debate between realist and anti-realist is whether the adequacy of Einstein's theory of the photon does require that photons be real.

Cosmic accidents

The simple inference argument considers just one theory, one phenomenon and one kind of entity. The cosmic accident argument notes that often in the growth of knowledge a good theory will explain diverse phenomena which had not hitherto been thought of as connected. Conversely, we often come at the same brute entities by quite different modes of reasoning. Hans Reichenbach called this the common cause argument, and it has been revived by Wesley Salmon.[3] His favoured example is not the photoelectric effect but another of Einstein's triumphs. In 1905 Einstein also explained the Brownian movement – the way in which, as we now say, pollen particles are bounced around in a random way by being hit by molecules in motion. When Einstein's calculations are combined

2 J.J.C. Smart, 'Difficulties for realism in the philosophy of science', in *Logic, Methodology and Philosophy of Science VI*, Proceedings of the 6th International Congress of Logic, Methodology and Philosophy of Science, Hannover, 1979, pp. 363–75.
3 Wesley Salmon, 'Why ask, "Why?" An Inquiry Concerning Scientific Explanation', *Proceedings and Addresses of the American Philosophical Association* 51 (1978), pp. 683–705.

with the results of careful experimenters, we are able, for example, to compute Avogadro's number, the number of molecules of an arbitrary gas contained in a given volume at a set temperature and pressure. This number had been computed from numerous quite different sources ever since 1815. What is remarkable is that we always get essentially the same number, coming at it from different routes. The only explanation must be that there *are* molecules, indeed, some 6.023×10^{23} molecules per gram-mole of any gas.

Once again, this seems to me to beg that realist/anti-realist issue. The anti-realist agrees that the account, due to Einstein and others, of the mean free path of molecules is a triumph. It is empirically adequate – wonderfully so. The realist asks why is it empirically adequate – is that not because there just are molecules? The anti-realist retorts that explanation is no hall-mark of truth, and that all our evidence points only to empirical adequacy. In short the argument goes around in circles (as, I contend, do all arguments conducted at this level of discussion of theories).

The success story

The previous considerations bear more on the existence of entities; now we consider the truth of theories. We reflect not on one bit of science but on 'Science' which, Hilary Putnam tells us, is a Success. This is connected with the claim that Science is converging on the truth, as urged by many, including W. Newton-Smith in his book *Rationality* (1982). Why is Science Successful? It must be because we are converging on the truth. This issue has now been well aired, and I refer you to a number of recent discussions.[4] The claim that here we have an 'argument' drives me to the following additional expostulations:

1 The phenomenon of growth is at most a monotonic increase in knowledge, not convergence. This trivial observation is important, for 'convergence' implies somewhat that there is *one* thing being converged on, but 'increase' has no such implication. There can be heapings up of knowledge without there being any unity of

---

4 Among many arguments in favour of this idea of convergence, see R.N. Boyd, 'Scientific realism and naturalistic epistemology', in P.D. Asquith and R. Giere (eds.), *PSA 1980*, Volume 2, Philosophy of Science Assn., East Lansing, Mich., pp. 613–62, and W.H. Newton-Smith, *The Rationality of Science*, London, 1981. For a very powerful statement of the opposite point of view, see L. Laudan, 'A confutation of convergent realism', *Philosophy of Science* 48 (1981), pp. 19–49.

science to which they all add up. There can also be an increasing depth of understanding, and breadth of generalization, without anything properly called convergence. Twentieth-century physics is a witness to this.

2 There are numerous merely sociological explanations of the growth of knowledge, free of realist implications. Some of these deliberately turn the 'growth of knowledge' into a pretence. On Kuhn's analysis in *Structure*, when normal science is ticking over nicely, it is solving the puzzles that it creates as solvable, and so growth is built in. After revolutionary transition, the histories are rewritten so that early successes are sometimes ignored as uninteresting, while the 'interesting' is precisely what the post-cataclysmic science is good at. So the miraculously uniform growth is an artifact of instruction and textbooks.

3 What grows is not particularly the strictly increasing body of (nearly true) *theory*. Theory-minded philosophers fixate on accumulation of theoretical knowledge – a highly dubious claim. Several things do accumulate. (a) Phenomena accumulate. For example, Willis Lamb is trying to do optics without photons. Lamb may kill off the photons but the photoelectric effect will still be there. (b) Manipulative and technological skills accumulate – the photoelectric effect will still be opening the doors of supermarkets. (c) More interestingly to the philosopher, styles of scientific reasoning tend to accumulate. We have gradually accumulated a horde of methods, including the geometrical, the postulational, the model-building, the statistical, the hypothetico-deductive, the genetic, the evolutionary, and perhaps even the historicist. Certainly there is growth of types (a), (b), and (c), but in none of them is there any implication about the reality of theoretical entities or the truth of theories.

4 Perhaps there is a good idea, which I attribute to Imre Lakatos, and which is foreshadowed by Peirce and the pragmatism soon to be described. It is a route open to the post-Kantian, post-Hegelian, who has abandoned a correspondence theory of truth. One takes the growth of knowledge to be a given fact, and tries to characterize truth in terms of it. This is not explanation by assuming a reality, but a definition of reality as 'what we grow to'. That may be a mistake, but at least it has an initial cogency. I describe it in Chapter 8 below.

5 Moreover, there are genuine conjectural inferences to be drawn from the growth of knowledge. To cite Peirce again, our talents at forming roughly the right expectations about the human-sized world may be accounted for by the theory of evolution. If we regularly formed the wrong expectations, we would all be dead. But we seem to have an uncanny ability to formulate structures that explain and predict both the inner constitution of nature, and the most distant realms of cosmology. What can it have benefited us, in terms of survival, that we have a brain so tooled for the lesser and the larger universe? Perhaps we should guess that people are indeed rational animals that live in a rational universe. Peirce made a more instructive if implausible proposal. He asserted that strict materialism and necessitarianism are false. The whole world is what he called 'effete mind', which is forming habits. The habits of inference that we form about the world are formed according to the same habits that the world used as it acquired its increased spectrum of regularities. That is a bizarre and fascinating metaphysical conjecture that might be turned into an explanation of 'the success of science'.

How Peirce's imagination contrasts with the banal emptiness of the Success Story or convergence argument for realism! Popper, I think, is a wiser self-professed realist than most when he writes that it never makes sense to ask for the explanation of our success. We can only have the faith to hope that it will continue. If you must have an explanation of the success of science, then say what Aristotle did, that we are rational animals that live in a rational universe.

4   Pragmatism

Pragmatism is the American philosophy founded by Charles Sanders Peirce (1839–1914), and made popular by William James (1842–1910). Peirce was a cantankerous genius who obtained some employment in the Harvard Observatory and the US Coast and Geodesic survey, both thanks to his father, then one of the few distinguished mathematicians in America. In an era when philosophers were turning into professors, James got him a job at Johns Hopkins University. He created a stir there by public misbehaviour (such as throwing a brick at a ladyfriend in the street), so the President of the University abolished the whole Philosophy Department, then created a new department and hired everyone back – except Peirce. Peirce did not like James's popularization of pragmatism, so he invented a new name for his ideas – pragmaticism – a name ugly enough, he would say, that no one would steal it. The relationship of pragmaticism to reality is well stated in his widely reprinted essay, 'Some consequences of four incapacities' (1868).

And what do we mean by the real? It is a conception which we must first have had when we discovered that there was an unreal, an illusion; that is, when we first corrected ourselves. . . . *The real, then, is that which, sooner or later, information and reasoning would finally result in*, and which is therefore independent of the vagaries of me and you. Thus, the very origin of the conception of reality shows that this conception essentially involves the notion of a COMMUNITY, without definite limits, and capable of a definite increase of knowledge. And so those two series of cognition – the real and the unreal – consist of those which, at a time sufficiently future, the community will always continue to reaffirm; and of those which, under the same conditions, will ever after be denied. Now, a proposition whose falsity can never be discovered, and the error of which therefore is absolutely incognizable, contains, upon our principle, absolutely no error. Consequently, that which is thought in these cognitions is the real, as it really is. There is nothing, then, to prevent our knowing outward things as they really are, and it is most likely that we do thus know them in numberless cases, although we can never be absolutely certain of doing so in any special case. (*The Philosophy of Peirce*, J. Buchler (ed.), pp. 247f.)

Precisely this notion is revived in our day by Hilary Putnam, whose 'internal realism' is the topic of Chapter 7.

The road to Peirce

Peirce and Nietzsche are the two most memorable philosophers writing a century ago. Both are the heirs of Kant and Hegel. They represent alternative ways to respond to those philosophers. Both took for granted that Kant had shown that truth cannot consist in some correspondence to external reality. Both took for granted that process and possibly progress are essential characteristics of the nature of human knowledge. They had learned that from Hegel.

Nietzsche wonderfully recalls how the true world became a fable. An aphorism in his book, *The Twilight of the Idols*, starts from Plato's 'true world – attainable for the sage, the virtuous man'. We arrive, with Kant, at something 'elusive, pale, Nordic, Königsbergian'. Then comes Zarathrustra's strange semblance of subjectivism. That is not the only post-Kantian route. Peirce tried to replace truth by method. Truth is whatever is in the end delivered to the community of inquirers who pursue a certain end in a certain way.

Thus Peirce is finding an objective substitute for the idea that truth is correspondence to a mind-independent reality. He sometimes called his philosophy objective idealism. He is much impressed with the need for people to attain a stable set of beliefs. In a famous essay on the fixation of belief, he considers with genuine seriousness the notion that we might fix our beliefs by following authority, or by believing whatever first comes into our heads and sticking to it. Modern readers often have trouble with this essay, because they do not for a moment take seriously that Peirce held an Established (and powerful) Church to be a very good way to fix beliefs. If there is nothing to which true belief has to correspond, why not have a Church fix your beliefs? It can be very comforting to know that your Party has the truth. Peirce rejects this possibility because he holds as a fact of human nature (not of pre-human truth) that there will in the end always be dissidents. So you want a way to fix beliefs that will fit in with this human trait. If you can have a method which is internally self-stabilizing, which acknowledges permanent fallibility and yet at the same time tends to settle down, then you will have found a better way to fix belief.

Repeated measurements as the model of reasoning

Peirce is perhaps the only philosopher of modern times who was quite a good experimenter. He made many measurements, including a determination of the gravitational constant. He wrote extensively on the theory of error. Thus he was familiar with the way in which a sequence of measurements can settle down to one basic value. Measurement, in his experience, converges, and what it converges on is by definition correct. He thought that all human beliefs would be like that too. Inquiry continued long enough would lead to a stable opinion about any issue we could address. Peirce did not think that truth is correspondence to the facts: the truths are the stable conclusions reached by that unending COMMUNITY of inquirers.

This proposal to substitute method for truth – which would still warrant scientific objectivity – has all of a sudden become popular again. I think that it is the core of the methodology of research programmes of Imre Lakatos, and explained in Chapter 8. Unlike Peirce, Lakatos attends to the motley of scientific practices and so does not have the simplistic picture of knowledge settling down by a repeated and slightly mindless process of trial and error. More recently Hilary Putnam has become Peircian. Putnam does not think that Peirce's account of the method of inquiry is the last word, nor does he propose that there is a last word. He does think that there is an evolving notion of rational investigating, and that the truth is what would result from the results to which such investigation tends. In Putnam there is a double limiting process. For Peirce, there was one method of inquiry, based on deduction, induction, and, to some small degree, inference to the best explanation. Truth was, roughly, whatever hypothesizing, inducing, and testing settled down upon. That is one limiting process. For Putnam the methods of inquiry can themselves grow, and new styles of reasoning can build on old ones. But he hopes that there will be some sort of accumulation here, rather than abrupt displacement of one style of reasoning just replacing another one. There can then be two limiting processes: the long term settling into a 'rationality' of accumulated modes of thinking, and the long term settling into facts that are agreed to by these evolving kinds of reason.

Vision

Peirce wrote on the whole gamut of philosophical topics. He has gathered about him a number of coteries who hardly speak to each other. Some regard him as a predecessor of Karl Popper, for nowhere else do we find so trenchant a view of the self-correcting method of science. Logicians find that he had many premonitions of how modern logic would develop. Students of probability and induction rightly see that Peirce had as deep an understanding of probabilistic reasoning as was possible in his day. Pierce wrote a great deal of rather obscure but fascinating material on signs, and a whole discipline that calls itself semiotics reveres him as a founding father. I think him important because of his bizarre proposal that one just is one's language, a proposal that has become a centrepiece of modern philosophy. I think him important because he was the first person to articulate the idea that we live in a universe of chance, chance that is both indeterministic, but which because of the laws of probability accounts for our false conviction that nature is governed by regular laws. A glance at the index at the end of this book will refer you to other things that we can learn from Peirce. Peirce has suffered from readers of narrow vision, so he is praised for having had this precise thought in logic, or that inscrutable idea about signs. We should instead see him as a wild man, one of the handful who understood the philosophical events of his century and set out to cast his stamp upon them. He did not succeed. He finished almost nothing, but he began almost everything.

The branching of the ways

Peirce emphasized rational method and the community of inquirers who would gradually settle down to a form of belief. Truth is whatever in the end results. The two other great pragmatists, William James and John Dewey, had very different instincts. They lived, if not for the now, at least for the near future. They scarcely addressed the question of what might come out in the end, if there is one. Truth is whatever answers to our present needs, or at least those needs that lie to hand. The needs may be deep and various, as attested in James's fine lectures, *The Varieties of Religious Experience*. Dewey gave us the idea that truth is warranted acceptability. He thought of language as an instrument that we use to mould our experiences to suit our ends. Thus the world, and our representation of it, seems to become at the hands of Dewey very much of a social construct. Dewey despised all dualisms – mind/matter, theory/practice, thought/action, fact/value. He made fun of what he called the spectator theory of knowledge. He said it resulted from the existence of a leisure class, who thought and wrote philosophy, as opposed to a class of entrepreneurs and workers, who had not the time for just looking. My own view, that realism is more a matter of intervention in the world, than of representing it in words and thought, surely owes much to Dewey.

There is, however, in James and Dewey, an indifference to the Peircian vision of inquiry. They did not care what beliefs we settle on in the long run. The final human fixation of belief seemed to them a chimaera. That is partly why James's rewriting of pragmatism was resisted by Peirce. This same disagreement is enacted at the very moment. Hilary Putnam is today's Peircian. Richard Rorty, in his book *Philosophy and the Mirror of Nature* (1979), plays some of the parts acted by James and Dewey. He explicitly says that recent history of American philosophy has got its emphases wrong. Where Peirce has been praised, it has been only for small things. (My section above on Peirce's vision, obviously disagrees.) Dewey and James are the true teachers, and Dewey ranks with Heidegger and Wittgenstein as the three greats of the twentieth century. However Rorty does not write only to admire. He has no Peirce/Putnam interest in the long run nor in growing canons of rationality. Nothing is more reasonable than anything else, in the long run. James was right. Reason is whatever goes in the conversation of our days, and that is good enough. It may be sublime, because of what it inspires within us and among us. There is nothing that makes one conversation intrinsically more rational than another. Rationality is extrinsic: it is whatever we agree on. If there is less persistence among fashionable literary theories than among fashionable chemical theories, that is a matter of sociology. It is not a sign that chemistry has a better method, nor that it is nearer to the truth.

Thus pragmatism branches: there are Peirce and Putnam on the one hand, and James, Dewey and Rorty on the other. Both are anti-realist, but in somewhat different ways. Peirce and Putnam optimistically hope that there is something that sooner or later, information and reasoning would finally result in. That, for them, is the real and the true. It *is* interesting for Peirce and Putnam both to define the real and to know what, within our scheme of things, will pan out as real. This is not of much interest to the other sort of pragmatism. How to live and talk is what matters, in those quarters. There is not only no external truth, but there are no external or even evolving canons of rationality. Rorty's version of pragmatism is yet another language-based philosophy, which regards all our life as a matter of conversation. Dewey rightly despised the spectator theory of knowledge. What might he have thought of science as conversation? In my opinion, the right track in Dewey is the attempt to destroy the conception of knowledge and reality as a matter of thought and of representation. He should have turned the minds of philosophers to experimental science, but instead his new followers praise talk.

Dewey distinguished his philosophy from that of earlier philosophical pragmatists by calling it *instrumentalism*. This partly indicated the way in which, in his opinion, things we make (including all tools, including language as a tool) are instruments that intervene when we turn our experiences into thoughts and deeds that serve our purposes. But soon 'instrumentalism' came to denote a philosophy of science. An instrumentalist, in the parlance of most modern philosophers, is a particular kind of anti-realist about science – one who holds that theories are tools or calculating devices for organizing descriptions of phenomena, and for drawing inferences from past to future. Theories and laws have no truth in themselves. They are only instruments, not to be understood as literal assertions. Terms that seemingly denote invisible entities do not function as referential terms at all. Thus instrumentalism is to be contrasted with van Fraassen's view, that theoretical expressions are to be taken literally – but not believed, merely 'accepted' and used.

How do positivism and pragmatism differ?

The differences arise from the roots. Pragmatism is an Hegelian doctrine which puts all its faith in the process of knowledge. Positivism results from the conception that seeing is believing. The pragmatist claims no quarrel with common sense: surely chairs and electrons are equally real, if indeed we shall never again come to doubt their value to us. The positivist says electrons cannot be believed in, because they can never be seen. So it goes through all the positivist litany. Where the positivist denies causation and explanation, the pragmatist, at least in the Peircian tradition, gladly accepts them – so long as they turn out to be both useful and enduring for future inquirers.

5   Incommensurability

Why is so shop-soiled a topic as scientific realism once again prominent in the philosophy of science? Realism fought a great battle when Copernican and Ptolemaic world views were at issue long ago. Towards the end of the nineteenth century worries about atomism strongly contributed to anti-realism among philosophers of science. Is there a comparable scientific issue today? Maybe. One way to understand quantum mechanics is to take an idealist line. Some people argue that human observation plays an integral role in the very nature of a physical system, so that the system changes simply when it is measured. Talk of 'the measurement problem in quantum mechanics', the 'ignorance interpretation', and 'the collapse of the wave packet' make it no accident that contributions to the philosophy of quantum mechanics play an important part in the writings of the more original figures in the realist debate. A number of the ideas of Hilary Putnam, Bas van Fraassen or Nancy Cartwright seem to result from taking quantum mechanics as the model of all science.

Conversely, numerous physicists wax philosophical. Bernard d'Espagnat has made one of the most important recent contributions to a new realism. He is partly motivated by the dissolution, in some parts of modern physics, of old realist concepts such as matter and entity. He is especially driven by some recent results, that bear the general name of Bell's inequality, and which have been thought to call in question concepts as various as logic, the temporal order of causation, and action at a distance. In the end he defends a realism different from any discussed in this book.

There are, then, problems within science that spur present thinking about realism. But problems of a particular science are never the whole story of a philosophical disturbance. Notoriously the Ptolemy/Copernicus debate that climaxed in the condemnation of Galileo had roots in religion. It involved our conception of the status of humanity in the universe: are we at its centre or on the periphery? Anti-realist anti-atomism was part of late-nineteenth-century positivism. Likewise, in our time, Kuhn's historico-philosophical work has been a major element in the rediscussion of realism. It is not that he single-handedly wrought a transformation in the history and philosophy of science. When his book *The Structure of Scientific Revolutions* came out in 1962, similar themes were being expressed by a number of voices. Moreover a new discipline, the history of science, was forming itself. In 1950 it was mostly the province of gifted amateurs. By 1980 it was an industry. Young Kuhn, training as a physicist, was attracted to history just at the moment when many other people were looking that way. As I have said in my Introduction, the fundamental transformation in philosophical perspective was this: science became an historical phenomenon.

This revolution had two interconnected effects on philosophers. There was the crisis of rationality that I described. There was also a great wave of doubt about scientific realism. With each paradigm shift, we come, so Kuhn hints, to see the world differently – perhaps we live in a different world. Nor are we converging on a true picture of the world, for there is none to be had. There is no progress towards the truth, but only increased technology and perhaps progress 'away from' ideas that we shall never again find tempting. Is there then a real world at all?

Within this family of ideas one catchword has had a special vogue – *incommensurability*. It has been said that successive and competing theories within the same domain 'speak different languages'. They cannot strictly be compared to each other nor translated into each other. The languages of different theories are the linguistic counterparts of the different worlds we may inhabit. We can pass from one world or one language to another by a gestalt-switch, but not by any process of understanding.

The realist about theories cannot welcome this view, in which the aim of discovering the truth about the world is dispersed. Nor is the realist about entities pleased, for all theoretical entities seem totally theory-bound. There may be electrons within our present theory, but no sense is left for the claim that there just are electrons, regardless of what we think. There have been numerous theories about electrons professed by distinguished scientists: R.A. Millikan, H.A. Lorentz, and Niels Bohr had very different ideas. The incommensurabilist says that they meant something different, in each case, by the word 'electron'. They were talking about different things, says the incommensurabilist, whereas the realist about entities thinks they were talking about electrons.

Hence, although incommensurability is an important topic for discussions of rationality, it also opposes scientific realism. A little care, however, makes it seem less of a dragon than is sometimes supposed.

Kinds of incommensurability

The new philosophical use of the word 'incommensurable' is the product of conversations between Paul Feyerabend and Thomas Kuhn on Berkeley's Telegraph Avenue around 1960. What did it mean before these two men refashioned it? It has an exact sense in Greek mathematics. It means 'no common measure'. Two lengths have common measure if you can lay m of the first lengths against exactly n of the second, and thus measure the one by the other. Not all lengths are commensurable. The diagonal on a square has no common measure with the length of the sides, or, as we now express this fact, $\sqrt{2}$ is not a rational fraction, m/n.

Philosophers have nothing so precise in mind when they use the metaphor of incommensurability. They are thinking of comparing scientific theories, but of course there could be no *exact* measure for that purpose. After twenty years of heated debate the very word 'incommensurable' seems to point to three distinguishable things. I shall call them *topic-incommensurability*, *dissociation*, and *meaning-incommensurability*. The first two may be fairly straightforward but the third is not.

Accumulation and subsumption

Ernest Nagel's *The Structure of Science* of 1961 was a classic statement of much philosophy of science that had recently been written in English. (Titles can say so much. The hit of 1962 was *The Structure of Scientific Revolutions*.) Nagel tells of stable structures and continuity. He took for granted that knowledge tends to accumulate. From time to time one theory T is replaced by a successor T^*. When is it rational to switch theories? Nagel's idea was that the new T^* ought to be able to explain the phenomena that T explains, and it should also make whatever true predictions are made by T. In addition, it should either exclude some part of T that is erroneous, or cover a wider range of phenomena and predictions. Ideally $T^\star$ does both. In that case $T^\star$ *subsumes* T.

When $T^\star$ subsumes T there is, loosely speaking, a common measure for comparing the two; at any rate, the correct part of T is included in $T^\star$. So we might, by metaphor, say that T and $T^\star$ are commensurable. This very commensurability provides a basis for the rational comparison of theories.

Topic-incommensurability

Feyerabend and Kuhn made it clear that Nagel did not exhaust the possibilities for theory change. A successor theory may attack different problems, use new concepts and have applications different from the old theory. It may simply forget many former successes. The ways in which it recognizes, classifies, and above all produces phenomena may not match up well with the older account. For example, the oxygen theory of burning and bleaching did not at first apply to all the phenomena that fitted nicely into phlogiston. As an historical fact it was just not true that the new theory subsumed the old one.

In Nagel's opinion $T^\star$ ought to cover the same topics as T, and cover them at least as well as T; it should also cover some new topics. Such a sharing and extension of topics makes for commensurability between T and $T^\star$. Kuhn and Feyerabend said that often there is a radical shift in topics. We cannot say that successor $T^\star$ does the same job better than T, because they do different jobs.

Kuhn's picture of normal science, crisis, revolution, normal science makes such topic-incommensurability quite plausible. A crisis occurs in T when a family of counterexamples attracts widespread attention, but refuses to yield to revisions in T. A revolution redescribes the counterexamples, and produces a theory, that explains previously troublesome phenomena. The revolution succeeds if the new concepts resolve certain old problems and produces new approaches and topics to investigate. The resulting normal science may ignore a lot of triumphs of the preceding normal science. Hence although there will be some overlap between $T^\star$ and T, there may be nothing like Nagel's picture of subsumption. Moreover even where there is overlap, the ways in which $T^\star$ describes some phenomena may be so different from the description furnished by T that we may feel that these are not even understood in the same way.

In 1960, when most philosophers writing in English would have agreed with Nagel, Kuhn and Feyerabend came as a great shock. But by now topic-incommensurability by itself seems quite straightforward. It is an historical question whether the oxygen theory mostly moved on to a set of topics different from those studied by phlogiston. Doubtless there will be a great range of historical examples, starting on the one end from pure Nagelian subsumption, and arriving at the opposite extreme in which we wish to say that the successor theory totally replaced the topics, concepts and problems of T. In the extreme, students of a later generation educated on $T^\star$ may find T simply unintelligible until they play the role of historians and interpreters, relearning T from scratch.

Dissociation

A long enough time, and radical enough shifts in theory, may make earlier work unintelligible to a later scientific audience. Here it is important to make a distinction. An old theory may be forgotten, but still be intelligible to the modern reader who is willing to spend the time relearning it. On the other hand some theories indicate so radical a change that one requires something far harder than mere learning of a theory. Two examples suffice to make the contrast.

The five-volume *Celestial Mechanics* is a great Newtonian physics book written by Laplace around 1800. The modern student of applied mathematics can understand it. This is true even toward the end of the work where Laplace writes on caloric. Caloric is a substance, the substance of heat, and it is supposed to consist of small particles with a repulsive force that decays very rapidly with distance. Laplace is proud to solve some important problems with his caloric model. He is able to provide the first derivation of the speed of sound in air. Laplace gets roughly the observed velocity, while Newton's derivations gave quite the wrong answer. We no longer believe there is such a substance as caloric, and we have entirely replaced Laplace's theory of heat. But we can work it out and understand what he is doing.

For a contrast turn to the many volumes of Paracelsus, who died in 1541. He exemplifies a Northern European Renaissance tradition of a bundle of hermetic interests: medicine, physiology, alchemy, herbals, astrology, divination. Like many another 'doctor' of the day, he practised all of these as part of a single art. The historian can find in Paracelsus anticipations of later chemistry and medicine. The herbalist can retrieve some forgotten lore from his remarks. But if you try to read him you will find someone utterly different from us.

It is not that we cannot understand his words, one by one. He wrote in dog-Latin and proto-German, but that is no serious problem. He is now translated into modern German and some of his work is available in English. The tone is well suggested by passages like this: 'Nature works through other things, such as pictures, stones, herbs, words, or when she makes comets, similitudes, halos and other unnatural products of the heavens.' It is the ordering of thought that we cannot grasp here, for it is based on a whole system of categories that is hardly intelligible to us.

Even when we seem to be able to understand the words perfectly well, we are left in a fog. Many a Renaissance writer of high seriousness and intelligence makes extraordinary statements about the origins of ducks or geese or swans. Rotting logs floating in the Bay of Naples will generate geese. Ducks are generated from barnacles. People then knew all about ducks and geese: they had them in their barnyards nearby. Swans were kept in semi-cultivation by the ruling classes. What is the force of these absurd propositions about barnacles and logs? We do not lack sentences to express these thoughts. We have words, such as this one to be found alike in Johnson's *Dictionary* (1755) and the *Oxford English Dictionary*: '*Anatiferous* – producing ducks or geese, that is producing barnacles, formerly supposed to grow on trees, and, dropping off into the water below, to turn into tree-geese.' The definition is plain enough, but what is the point of the idea?

Paracelsus is not a closed book. One can learn to read him. One can even imitate him. There were in his day many imitations that we now call pseudo-Paracelsus. You could get sufficiently into his way of thinking to forge another volume of pseudo-Paracelsus. But to do that you would have to recreate an alien system of thought that we now only barely recall, for example, in homeopathic medicine. The trouble is not just that we think Paracelsus wrote falsely, but that we cannot attach truth or falsehood to a great many of his sentences.

His style of reasoning is alien. Syphilis is to be treated by a salve of mercury and by internal administration of the metal, because the metal mercury is the sign of the planet, Mercury, and that in turn signs the market place, and syphilis is contracted in the market place. Understanding this is an entirely different exercise from learning Laplace's theory of caloric.

Paracelsus's discourse is incommensurable with ours, because there is no way to match what he wanted to say against anything we want to say. We can express him in English, but we cannot assert or deny what is being said. At best one can start talking his way only if one becomes alienated or dissociated from the thought of our own time. Hence I shall say that the contrast between ourselves and Paracelsus is *dissociation*.

We do not strain a metaphor if we say that Paracelsus lived in a different world from ours. There are two strong linguistic correlates of dissociation. One is that numerous Paracelsan statements are not among our candidates for truth-or-falsehood. The other is that forgotten styles of reasoning are central to his thought. I argue elsewhere that these two aspects are closely connected. An interesting proposition is in general true-or-false only if there is a style of reasoning that helps one settle its truth value.[1] Quine and others write of conceptual schemes, by which they mean a body of sentences held for true. That is, I think, a mistaken characterization. A conceptual scheme is a network of possibilities, whose linguistic formulation is a class of sentences up for grabs as true or false. Paracelsus viewed the world as a different network of possibilities, embedded in different styles of reasoning from ours, and that is why we are dissociated from him.

Although Paul Feyerabend has spoken of incommensurability in many domains of science, his mature thoughts in *Against Method* are mostly about what I call dissociation. His prize example is the shift from archaic to classical Greece. Drawing chiefly on epic poetry and paintings on urns, he contends that Homeric Greeks literally saw things differently from Athenians. Whether or not this is correct, it is a much less surprising claim than one that says, for example, that each cohort of physicists has been referring to different things, when speaking of electrons.

---

1 See I. Hacking, 'Language, truth and reason', in M. Hollis and S. Lukes (eds.), *Rationality and Relativism*, Oxford, 1982, pp. 48–66.

Many examples lie between the extremes of Laplace and Paracelsus. The historian soon learns that old texts constantly conceal from us the extent to which they are dissociated from our ways of thought. Kuhn tells us, for example, that Aristotle's physics relies on ideas of motion that are dissociated from ours, and one can understand him only by recognizing the network of his words. Kuhn is one of many historians to teach the need to rethink the works of our predecessors in their way, not ours.

Meaning-incommensurability

The third kind of incommensurability is not historical but philosophical. It starts from asking about the meaning of terms that stand for theoretical, unobservable entities.

How do names for theoretical entities or processes get their meaning? We may have the idea that a child could grasp the use of words like 'hand' and 'sick' and 'sad' and 'horrible' by being shown things to which these words apply (including his own hands, his own sadness). Whatever be our theory of language acquisition, the manifest presence or absence of hands or sadness must be a help in catching on to what the words mean. But theoretical terms refer – almost by definition – to what cannot be observed. How do they get their meaning?

We can give some meanings by definitions. But in the case of deep theories, any definition would itself involve other theoretical terms. Moreover we seldom use definitions for starting an understanding. We explain theoretical terms by talking theory. This has long suggested that the sense of the terms is given by a string of words from the theory itself. The meaning of individual terms in the theory is given by their position within the structure of the entire theory.

On this view of meaning, it would follow that 'mass' in Newtonian theory would not mean the same 'mass' in relativistic mechanics. 'Planet' in Copernican theory will not mean the same as 'planet' in Ptolemaic theory, and indeed the sun is a planet for Ptolemy but not for Copernicus. Such conclusions are not necessarily problematic. Did not the sun itself mean something different when Copernicus put it at the centre of our system of planets? Why should it matter if we say that 'planet' or 'mass' evolved new meanings as people thought more about planets and mass? Why should we fuss about meaning change? Because it seems to matter when we start comparing theories.

Let *s* be a sentence about mass, asserted by relativistic mechanics and denied by Newtonian mechanics. If the word 'mass' gets its meaning from its place in a theory, it will mean something different depending on whether it is used in Newtonian or relativistic mechanics. Hence the sentence *s*, asserted by Einstein, must differ in meaning from the sentence *s* denied by Newton. Indeed, let *r* be another sentence using the word 'mass', but which unlike *s*, is asserted by both Newton and Einstein. We cannot say that the sentence *r*, which occurs in the Newtonian theory, is subsumed in the relativistic theory. For 'mass' will not mean the same in both contexts. There will be no one proposition, the shared meaning of *r*, which is common to both Newton and Einstein.

That is incommensurability with a vengeance. There is no common measure for any two theories that employ theoretical terminology because in principle they can never discuss the same issues. There cannot be theoretical propositions that one theory shares with its successor. Nagel's doctrine of subsumption then becomes logically impossible, simply because what *T* says cannot even be asserted (or denied) in the successor theory *T**. Such are the remarkable claims for meaning-incommensurability. One can even begin to wonder whether crucial experiments are logically possible. If an experiment is to decide between theories, would there not have to be a sentence asserting what one theory predicts and what the other denies? Can there be such a sentence?

The doctrine of meaning-incommensurability was met by cries of outrage. The whole idea was said to be incoherent. For example: no one would deny that astronomy and genetics are incommensurable – they are about different domains. But meaning-incommensurability says that competing or successive theories are incommensurable. How could we even call them competing or successive if we did not recognize them to be about the same subjects, and hence be making a comparison between them? There are other equally shallow responses to meaning-incommensurability. Then there are deep ones, of which the best is Donald Davidson's. Davidson implies that incommensurability makes no sense, because it rests on the idea of different and incomparable conceptual schemes. But, he urges, the very idea of a conceptual scheme is incoherent.[2]

At a more straightforward level it has been carefully argued, for example by Dudley Shapere, that there is enough sameness of meaning between successive theories to allow for theory comparison.[3] Shapere is among those, now including Feyerabend, who suppose that such matters are best discussed without bringing in the idea of meaning at all. I agree. But at the root of meaning-incommensurability is a question about how terms denoting theoretical entities get their meaning. The question presupposes a rough conception of meaning. Given that the question has been raised and such a storm provoked, we are obliged to produce a better rough conception of meaning. Hilary Putnam has honoured that obligation, and we now turn to his theory of reference in order to evade meaning-incommensurability altogether.

2 D. Davidson, 'On the very idea of a conceptual scheme', *Proceedings and Addresses of the American Philosophical Association* 57 (1974), pp. 5–20.
3 D. Shapere, 'Meaning and scientific change', in R. Colodny (ed.), *Mind and Cosmos: Essays in Contemporary Science and Philosophy*, Pittsburgh, 1966, pp. 41–85.

6 Reference

If only philosophers of science had never troubled themselves about meaning we should have no doctrine of meaning-incommensurability. As it is, we need an alternative account of meaning which allows that people holding competing or successive theories may still be talking about the same thing. The most viable alternative is Hilary Putnam's.[1] He intended it as a part of his former scientific realism. He has since become increasingly anti-realist, but that is a story I reserve for the next chapter. For the present consider his meaning of 'meaning'.

Sense and reference

The word 'meaning' has many uses, many of which are more evocative than precise. Even if we stick to the commonplace meaning of words, as opposed to poems, there are at least two distinct kinds of meaning. They are distinguished in a famous 1892 essay by Gottlob Frege, 'On sense and reference'.

Consider two different kinds of answer to the question, What do you mean? Suppose I have just told you that the glyptodon brought by Richard Owen from Buenos Aires has now been restored. Most people do not know the meaning of the word 'glyptodon' and so may ask, What do you mean?

If we are standing in the museum I may simply point to a largish and preposterously shaped skeleton. *That* is what I mean. In Frege's parlance, that very skeleton is the reference of my words, 'The glyptodon brought by Richard Owen from Buenos Aires.'

On the other hand, since you probably do not have a clue what the word 'glyptodon' means, I may tell you that a glyptodon is an enormous, extinct South American mammal akin to the armadillo, but with fluted teeth. With this definition I indicate what Frege would have called the *sense* of the word 'glyptodon'.

---

1 All references to Hilary Putnam are to 'The meaning of "meaning"' and other essays reprinted in Volume 2 of his *Philosophical Papers. Mind, Language and Reality*, Cambridge, 1979.

It is natural to think of a phrase as having a sense, namely what we understand by it, that enables us to pick out the reference, if there is one. Hearing the definition of 'glyptodon' I can go to a museum and try to find their skeletons, if any, without looking at the labels beneath the specimens. Frege thought that a word has a standard sense, which is what makes a scientific tradition possible. The sense is what is shared by all communicators, and may be passed down from generation to generation of students.

Sense and meaning-incommensurability

Frege would have despised meaning-incommensurability but his way of looking at things helped lead into that trap. He taught us that an expression should have a definite fixed sense, which we apprehend, and which enables us to pick out the reference. Now add to this the unFregeian idea that we can grasp the sense of theoretical terms only by considering their place in a network of theoretical propositions. It seems to follow that the sense of such a term must change as the theory undergoes change.

We can evade this conclusion in several ways. One is to avoid breaking up meaning into just two components, sense and reference, with all the work being done by abstract, objective, senses. After all, the idea of meaning does not come in two nice packages that nature has labelled sense and reference. The sorting and the wrapping is the work of logicians and linguists. J.S. Mill did it in a slightly different way (connotation and denotation). So did the scholastic grammarians (intension and extension). French writers following the linguist Ferdinand de Saussure have a quite different split (signifier and signified). We may loosen Frege's strings and tie up the parcels differently. Doubtless there are many ways to do so. Hilary Putnam's is especially useful because, unlike all the other writers, he does not have just a pair of components of 'meaning'.

Putnam's meaning of 'meaning'

Dictionaries are mines of information. They do not state only abstract Fregeian senses, omitting all empirical non-linguistic facts about the world. Open one at random, and you'll learn, say, that the French gold coin, the Louis d'or, was first struck in 1640, and continued up to the Revolution. You'll learn that ancient Egyptian and Hindu religious art includes ritual representation of a water lily called a lotus – and that the fruit of the mythical lotus tree is held to produce dreamy contentment. A dictionary begins an entry with some pronunciation and grammar, proceeds past etymology to a lot of information, and may conclude with examples of usage. My concise dictionary ends the entry for 'it' with the example: 'It's a dirty business, this meat-canning.'

Putnam builds his account of meaning from an analogous string of components. We may think of him as leading a back-to-the-dictionary movement. I shall use two words as examples. One is his own choice, 'water', and the other is our word, 'glyptodon'.

Putnam's first component of meaning is grammatical. He calls it a *syntactic marker*. 'Glyptodon' is a count noun, and 'water' is a mass noun. That has to do with for example the formation of plurals. We say there is some water in the pit, but either, there is a glyptodon in the pit, or, there are some glyptodons in the pit. The words have different grammars. Putnam would also include among the syntactic markers indications that both words are concrete (as opposed to abstract) names.

Putnam's second component is a *semantic marker*. In our cases this will show the category of items to which the words apply. Both 'water' and 'glyptodon' are names of things found in nature, so Putnam enters 'natural kind term' among the semantic markers. Under 'water' he lists 'liquid'. Under 'glyptodon', he would put 'mammal'.

Stereotypes

Putnam's more original contribution is the third component, the *stereotype*. A stereotype is a conventional idea associated with a word, which might well be inaccurate. To use his example, a person who understands the word 'tiger' in our community must know that tigers are thought of as striped. Illustrations in children's books emphasize the stripiness of tigers; that is important for showing them to be pictures of *tigers*. Even if one thought that being striped is a sort of accident, and that tigers will soon adapt to the destruction of their forests by becoming a uniform desert-tan colour, it is still true that our standard tigers are striped. You need to know that to communicate at length about tigers. But it is not a self-contradiction to speak of a tiger that has lost its stripes. An entirely white tiger has been authentically recorded. Likewise, it is part of the sterotype of dogs that they are four-legged, even though my dog Bear has only three legs.

As part of the stereotype for 'water' Putnam gives us colourless, transparent, tasteless, thirst-quenching, etc. Under 'glyptodon' we might have enormous, extinct, South American, akin to the armadillo, with fluted teeth.

Notice that some of these elements may be mistaken. The word 'glyptodon' comes from the Greek for flute + tooth. It was invented by the paleontologist who discovered glyptodon remains in 1839, Richard Owen. But maybe the eponymous fluted teeth are a feature of only some glyptodons. Every single element in the stereotype could be wrong. Maybe we shall find small glyptodons. There were glyptodons in North America too. Perhaps the species is not extinct, but survives far up the Amazon or the Andes. Maybe Owen was wrong about the evolutionary tree, and the animal is not akin to the armadillo.

Likewise, we may add things to the stereotype. Glyptodons lived in the Pleistocene era. They had spiky tails with knobs on the end that could be used as clubs. They ate anything they could get their fluted teeth into. I have noticed that reference books written 70 years ago emphasize quite different features of the glyptodon from what I find today.

The division of linguistic labour

The elements of Putnam's stereotypes are not permanent criteria for the use of the word in question. A person may know the meaning of the word, and know how to use it in many situations, without knowing the present best criteria for the application of the word. I may know how to tell a glyptodon skeleton when I see one, but not be up on the criteria current among paleontologists. Putnam speaks of the division of linguistic labour. We rely on experts to know the best criteria and how to apply them. That kind of expertise is not a matter of knowing the meaning but of knowing the world.

Putnam suggests something of a hierarchy in our understanding. It is similar to the one presented by Leibniz long ago, in his *Meditations Concerning Truth and Ideas* (1684).

In the worst state, a person may simply not know what a word means. Thus in one of his papers Putnam asserts that 'heather' is a synonym for 'gorse'. That is an innocent slip that charmingly illustrates Putnam's own distinctions. Gorse and heather are both plants characteristic of Scotland, for example, but gorse is a big shrub, spiky, with bright yellow flowers. Heather is low, soft, with tiny purple bell shaped flowers. Putnam must have not known, or forgotten, even the stereotypes for these shrubs. But it is doubtless a slip: he should have said that 'furze' is a synonym for 'gorse'. Fowler's *Modern English Usage* says that these two words are the rarest of pairs, perfect synonyms, used in the same regions interchangeably by the same speakers with no shade of difference in meaning.

Next, one may know what a word means and yet not be able to apply it correctly. Putnam, continuing his candid botanical confessions, tells us that he cannot tell a beech tree from an elm. Hence, he has what Leibniz called an *obscure* idea of a beech tree: in Leibniz's words, 'when my vague idea of a flower or animal which I once previously saw does not suffice for me to recognize a new instance when I encounter one'.

Next, one may be able to tell beech trees from elms, or to tell gold from other substances, without knowing the standard criteria or how to apply them. This is what Leibniz calls having a *clear* idea. One has a *distinct* idea when one knows the criteria and how to use them. Putnam and Leibniz use the same example: an assayer is an expert who knows the principles for distinguishing gold, and can apply the tests. The assayer has a distinct idea of gold.

Only a few experts have distinct ideas, that is, know the criteria appropriate in some domain. But in general we all know the meanings of the common words like 'gold' or 'beech' for which there do exist definite criteria. Perhaps these words would not have quite their present currency were there not experts in the offing. Putnam conjectures that the division of linguistic labour is an important part of any linguistic community. Note too that expert criteria may change. Assayers do not use the same techniques now as they did in the time of Leibniz. It is also common for the first stab at defining a species to flounder. Stereotypic features are recognized, but not enough is known about the things to know what is important. What then is constant in meaning? Putnam makes everything turn on reference and extension.

Reference and extension

The *reference* of a natural kind term is the natural kind in question – if indeed there is such a natural kind. The reference of 'water' is a certain kind of stuff, namely H_2O. The *extension* of a term is the set of things that it is true of. Thus the extension of the term glyptodon is the set of all past, present and future glyptodons. What if 'glyptodon' is not a natural kind? Imagine that the paleontologists made a terrible mistake, and all the fluted teeth were from one sort of animal, while the armadillo-like shell was from another. There never was a glyptodon. Then 'glyptodon' is not a natural kind term and the question of its extension does not arise. If it must arise, the extension is the empty set.

The Putnamian account of meaning differs from previous ones in that it includes the extension, or the reference (or both) as part of the meaning. These, and not Fregeian sense, are what are held constant from generation to generation.

The meaning of 'meaning'

What is the meaning of the word 'glyptodon'? Putnam's answer is a vector with four components: syntactic markers, semantic markers, stereotype, extension. In practice, then, we should have:

Glyptodon: [Concrete count noun]. [Names a natural kind, a mammal]. [Extinct, primarily South American, enormous, akin to the armadillo, has a gigantic solid shell up to five feet long with no movable rings or parts, lived during the Pleistocene era, ate anything]. [. . . . . .].

Here we have nothing other than an uglified dictionary entry, except for the final square brackets that cannot be filled in. We cannot put all the glyptodons on to the page of the dictionary. Nor can we put in the natural kind. Pictorial dictionaries do their best, because they give us a photograph of a real glyptodon skeleton, or a sketch of how a glyptodon must have looked. Let us call the final [. . . . . .] the *dots of extension*.

Reference and incommensurability

Stereotypes may change as we find out more about a certain kind of thing or stuff. If we do have a genuine natural kind term, the reference of the term will remain the same, even though stereotypical opinions of the kind may change. *Thus the fundamental principle of identity for a term shifts from Fregeian sense to Putnamian reference.*

Putnam has always objected to meaning-incommensurability. The meaning-incommensurabilist says, implausibly, that whenever a theory changes, we cease to be talking about the same thing. Putnam realistically replies, that's absurd. Of course we are talking about the same thing, namely, the stable extension of the term.

When Putnam developed his theory of reference he was still a scientific realist. Meaning-incommensurability is bad for scientific realism, so it behooved Putnam to develop a theory of meaning that avoided the pitfalls of incommensurability. That is a negative result. There is also a positive one. For example, van Fraassen is an anti-realist who, like myself, thinks that the theory of meaning should have very little place in the philosophy of science. Still, he does tease the realist, who is confident that there are electrons: 'Whose electron did Millikan observe; Lorentz's, Rutherford's, Bohr's or Schrödinger's?' (*The Scientific Image*, p. 214). Putnam's account of reference provides the realist with the obvious reply: Millikan measured the charge on the electron. Lorentz, Rutherford, Bohr, Schrödinger and Millikan were all talking about electrons. They had different theories about electrons. Different stereotypes of electrons have been in vogue but it is the reference that fixes the sameness of what we are talking about.

This reply goes one dangerous step beyond what has thus far been said. In the case of water and glyptodons, there appears to be a good way of hooking up words and the world. We can at least point to some of that stuff, water; we can point to, photograph, or reconstruct a skeleton from a member of that species, glyptodon. We cannot point to electrons. We must show how Putnam's theory works on theoretical entities.

In the next few sections I describe some real-life namings. One ought to have a sense of the odd things that happen in science, as opposed to the limited range of unimaginative events that populate science fiction. It is a defect of Putnam's essays that he favours fictions over facts. The facts reveal some flaws in Putnam's simplified meaning of 'meaning'. Yet he has relieved us of the pseudo-problem of meaning-incommensurability. We do not need any theory about names in order to name electrons. (I secretly hold, on philosophical grounds, that in principle there can be no complete, general, theory of meaning or of naming.) We need only be assured that an obviously false theory is not the only possible theory. Putnam has done that.

I should also warn of some optional extras that are sometimes added to Putnam's account. Putnam's ideas evolved at the time that Saul Kripke independently presented a remarkable set of lectures now published under the title *Naming and Necessity*. Kripke holds that when one succeeds in naming a natural kind of thing, a thing of that kind must, as part of its very essence, of its very nature, be that kind. This harks back to a philosophy due to Aristotle, called essentialism. According to Kripke, if water is in fact H_2O then water is necessarily H_2O. As a matter of metaphysical necessity, it cannot be anything else. Of course for all we know, it might be something else, but that is an epistemic matter. This essentialism is only accidentally connected with Putnam's meaning of 'meaning'. His references need not be 'essences'. D.H. Mellor has given strong reasons to resist that idea, at least in so far as concerns the philosophy of science.[2] (That is another instance of the need for philosophers of science to be chary of theories of meaning.) Despite the intrinsic interest of Kripke's ideas for students of logic they are *not* to be added here to my version of Putnam's notions.

Dubbing the electron

New natural kinds, such as electrons, are often the result of initial speculations which are gradually articulated into theory and experiment.

Putnam urges that it is not necessary to point to an instance of a natural kind in order to pick it out and name it. Moreover, pointing is never enough. It is a well-known claim, often attributed to Wittgenstein, that any amount of pointing at examples and calling them apples is consistent with several – or indefinitely many – ways of applying the word 'apple' thereafter. Moreover, no amount of defining apples precludes, in principle, the possibility that the rule for using the word 'apple' will branch in indefinitely many different ways – not to mention our curious metaphors, such as that bit of the human neck called Adam's apple, or the oak apple, a large hard ball

2 D.H. Mellor, 'Natural kinds', *British Journal for the Philosophy of Science* 28 (1977), pp. 299–312.

on Californian oaks built as the nest of a parasite. No matter how we may feel about this supposedly Wittgensteinian doctrine, it is at least clear that pointing is never enough. What pointing does do is to provide us with a causal, historical, connection between our word 'apple' and a certain kind of fruit, namely apples. That connection could be established in other ways, as is illustrated by the historical development of theory and experiment around the word 'electron'.

Putnam tells a story of Bohr and the electron. Bohr, according to Putnam, had a theory about electrons. It was not a strictly correct theory, but he did draw our attention to this natural kind of thing. We should, says Putnam, apply a sort of principle of charity. Putnam calls it the principle of the benefit of the doubt, or, as he playfully puts it, the benefit of the dubbed. We may have doubts about what Bohr was doing, but given his place in our historical tradition, we should allow that he was indeed talking about electrons, albeit with an inadequate theory.

As usual I prefer truth to science fiction. Bohr did not invent the word 'electron', but took over a standard usage. He speculated about an already quite well-understood particle. The correct story is as follows. 'Electron' was the name suggested in 1891 for the natural unit of electricity. Johnstone Stoney had been writing about such a natural unit as early as 1874, and he dubbed it 'electron' in 1891. In 1897 J.J. Thomson showed that cathode rays consist of what were then called 'ultratomic particles' bearing a minimal negative charge. These particles were for long called 'corpuscles' by Thomson, who rightly thought he had got hold of some ultimate stuff. He determined their mass. Meanwhile, Lorentz was elaborating a theory of a particle of minimum charge which he quickly called the electron. Around 1908 Millikan measured this charge. The theory of Lorentz and others was shown to tie in rather nicely with the experimental work.

In my opinion Johnstone Stoney was speculating when he said there is a minimum unit of electric charge. We give him the benefit of the doubt, or rather, the benefit of the dubber, for he made up the name. If you like he, too, was talking about electrons (does it matter?) I have no doubt, however, in connection with Thomson and Millikan. They were well on the way to establishing the reality of these charged ultratomic particles, by experimentally determining their mass and charge. Thomson did have a false picture of the atom, often called the pudding picture. His atom had electrons in it like currants in a British pudding. But the incommensurabilist would be out of his mind if he said that Thomson measured the mass of something other than the electron – our electron, Millikan's electron, Bohr's electron.

The electron provides a happy illustration of Putnam's view of reference. We know ever so much more about electrons than Thomson did. We have regularly found that speculations about electrons and experiment on electrons can be made to mesh. In the early 1920s an experiment by O. Stern and W. Gerlach suggested that electrons have angular momentum, and soon after, in 1925, S.A. Goudsmit and G.E. Uhlenbeck had the theory of electron spin. No one at present doubts that the electron is a natural kind of fundamental importance. Many now imagine that the electron is not charged with the minimum unit of electric charge. Quarks, it is conjectured, have a charge of $1/3\ e$, but this doesn't hurt the reality or genuineness of electrons. It means only that one bit of the long-lived stereotype must be revised.

Acids: bifurcating kinds

One of Putnam's earliest examples concerns acids. 'Acid' does not denote a theoretical entity, but is a natural kind term like 'water'. The incommensurabilist would say that we mean something different by the word 'acid' than did Lavoisier or Dalton around 1800. Our theories about acids have changed substantially, but, Putnam says, we are still talking about the same kind of stuff as those pioneers of the new chemistry.

Is Putnam correct? Certainly there is an important cluster of properties in the professional stereotype for acids: acids are substances that in water solution taste sour, and change the colour of indicators such as litmus paper. They react with many metals to form hydrogen, and react with bases to form salts.

Lavoisier and Dalton would agree completely with this stereotype. Lavoisier happened to have a false theory about such substances, for he thought every acid had oxygen in it. Indeed he defined acids in that way, but in 1810 Davy showed that was a mistake, because muriatic acid is just HCl, what we now call hydrochloric acid. But there is no doubt that Lavoisier and Davy were talking about the same stuff.

Unfortunately for Putnam's choice of example, acids are not quite such a success story as electrons. Everything went along fine until 1923. In that year J.N. Brønsted in Norway and T.M. Lowry in Britain produced one new definition of 'acid', while G.N. Lewis in the United States produced a different definition. Today there are two natural kinds: Brønsted–Lowry acids and Lewis acids. Naturally these two 'kinds' both include all the standard acids, but some substances are acids of only one of the two kinds.

A Brønsted–Lowry acid is a member of a species that has a tendency to lose a proton (while bases have a tendency to gain one). A Lewis acid belongs to a species that can accept an electron pair from a base by forming a chemical bond composed of a shared electron pair. The two definitions happen to agree about bases but not about acids, because typical Lewis acids do not contain protons, which are a precondition of being a Brønsted–Lowry acid. As I understand it, most chemists prefer the Brønsted–Lowry account for most purposes, because it appears to provide a more satisfactory explanation of many facets of acidity. On the other hand the Lewis account is used for some purposes and was originally motivated by certain analogies with the older phenomenal characteristics of acids. One authority writes: 'Numerous lengthy polemical exchanges have taken place regarding the relative merits of the Brønsted–Lowry and Lewis definitions of acids and bases. The difference is essentially one of nomenclature and has little scientific content'. Still, the philosopher of naming must ask if Lavoisier meant Brønsted–Lowry acids or Lewis acids when he spoke of acids. Obviously he meant neither. Must we now mean one or the other species? No, only for certain specialized purposes. I think this example is somewhat in the spirit of Putnam's approach to meaning. There is, however, a problem if we take him literally. The meaning of 'acid' in 1920 (i.e. before 1923) ought to have the dots of extension filled in. By Brønsted and Lowry? Or by Lewis? Since both schools of chemistry were in part enlarging the theory of acids, we could try 'all the things agreed to be acids in 1920, before the enlarging got under way'. But that is almost certainly *not* a natural kind! We could try the intersection of the two definitions, but I doubt that is a natural kind either. This example reminds us that the notion of meaning is ill-adapted to philosophy of science. We should worry about kinds of acids, not kinds of meaning.

Caloric: the nonentity

People talk about phlogiston when they want a non-existent natural kind. Caloric is more interesting. When Lavoisier had done down the phlogiston theory, he still needed some account of heat. This was provided by caloric. Just as with the word 'electron', we know exactly when a substance was dubbed as caloric. It did not happen in a casual way. In 1785 there was a French chemical dubbing commission which decreed what things should be called. Many substances have been so called since that day. One new name was *calorique*, a precise term to replace one sense of the old word *chaleur*. Caloric was supposed to have no (or imponderable?) mass, and to be the substance we call heat. Not everyone accepted the official French definition. British writers would speak scathingly of 'what the French persist in calling calorific when there is a perfectly good English word, namely fire'.

There is a tendency to regard stuff like caloric as simply stupid. That is a mistake. As I remarked in Chapter 5, it plays a real role in the final volume of Laplace's great *Celestial Mechanics*, and not as 'fire' either. Laplace was a great Newtonian, and in the *Optics* Newton had speculated that the fine structure of the universe is composed of particles with forces of attraction and repulsion. The rates of extinction of these forces would vary from case to case (the rate of extinction for gravitational force is as the square of the distance). Laplace postulated different rates of extinction for both the attraction and repulsion of caloric directed to other particles. From this he was able to solve one of the outstanding problems of the century. Newtonian physics had hitherto made a complete hash of explaining the velocity of sound in air. From his assumptions about caloric Laplace was able to get a reasonable figure, very close to available experimental determinations. Laplace was justly proud of his achievement. Yet even before he published, Rumford was convincing some people that there could not be such a thing as caloric.

Caloric may seem to be no problem for Putnam's meaning of 'meaning'. This is a rare occasion in which we can fill in the dots of extension. The extension is the empty set. But this is too simple. Remember that Putnam was trying to explain how we and Lavoisier could both be talking about acids. Most of the answer was provided by the dots of extension. What about caloric? The community of French revolutionary scientists – men like Berthollet, Lavoisier, Biot and Laplace – all had different theories about caloric. They were still able to talk to each other, and it seems to me they were talking about the same thing. The glib remark is, yes, the same thing, namely nothing. But these four great men were not talking about the same thing as their predecessors, who discussed phlogiston, also of zero extension. They were very glad to know that caloric is not phlogiston. Putnam's theory does not give a very good account of why 'caloric' has the same meaning for all these people: a meaning different from phlogiston. Their stereotypes for caloric were different from those for phlogiston – but not *that* different. Nor, on Putnam's theory, is it stereotypes that fix meaning. I think the lesson is that the language game of naming hypothetical entities can occasionally work well even if no real thing is being named.

Mesons and muons: how theories steal names from experiments

It is easier to give old examples than recent ones because many old examples have become general knowledge. But philosophy of science loses in richness by sticking to the past. So my concluding example will be a little more up to date, and correspondingly harder to understand. It illustrates a simple point. You can baptize x's with the new name N, and then it is decided that completely different things y are N. Some other name has to be found for x's. Namings need not stick; they can be stolen. Anybody who thinks that reference works by a causal and historical connection to the thing named ought to reflect on the following example.

A meson is a medium weight particle, heavier than an electron, lighter than a proton. There are many kinds of mesons. A muon is rather like an electron, but 207 times heavier. Mesons are very unstable. They decay into lighter mesons and muons, and then into electrons, neutrinos and photons. Muons decay into electrons and two kinds of neutrinos. Most muons come from meson decay. Since muons are charged, they have to lose the charge when decaying. They do this by ionization, that is, by knocking electrons off atoms. Since this dissipates little energy, muons are very penetrating. They occur in cosmic rays and are that part of the ray that can travel miles under the surface of the earth to be detected deep in a mineshaft.

The fundamental fact about these two kinds of entity has to do with forces and interactions. There are four kinds of forces in the universe: electromagnetic, gravitation, weak and strong. More explanation of the latter two will be given in Chapter 16. For the present, they are just suggestive names. Strong forces bind together electrons and protons in the atom, while weak forces can be illustrated by radioactive decay. Mesons have to do with strong forces, and were originally postulated to explain how the atom stays together. They enter into strong interactions. Muons enter only into weak interactions.

As quantum mechanics became applied to electrodynamics, about 1930, there arose quantum electrodynamics, or QED for short. It has since proven to be the best theory of the universe yet devised, applying over a far wider range of phenomena and sizes of entities than anything previously known. (Perhaps it is the fulfilment of Newton's dream in the *Optics*.) In the beginning, like all physics, it would make simplifying assumptions, for example, that the electron occupies a point. It was taken for granted that some of its equations would have singularities with no solution to a real physical problem, and that one would rectify this by various *ad hoc* approximations, for example, adding extra terms to an equation.

It was at first thought that the available QED did not apply to the very penetrating particles in cosmic rays. They must be highly energized electrons, and electrons with that much energy would produce a singularity in the equations of QED. No one was much worried by this, for physics is mostly a matter of such adjustments in equations.

In 1934, H.A. Bethe and W.H. Heitler derived an important consequence of QED. It is called the energy-loss formula and it applies to electrons. In 1936 two groups of workers (C.D. Anderson and S.H. Neddermeyer; J.C. Street and E.C. Stevenson), studying cosmic rays with cloud chambers, were able to show that the energetic particles in cosmic rays did not obey the Bethe–Heitler energy-loss formula. In fact at that time QED was confirmed, contrary to expectations. The equations of QED were fine; however there was a new particle hitherto undreamt of. This was named the mesotron, because its mass lay between the electron and proton. This name was soon shortened to meson.

Meanwhile in 1935 H. Yukawa had been speculating about what

holds the atom together. He postulated that there must be a new kind of object, also intermediate in mass between electron and proton. Evidently, he was addressing a problem entirely different from cosmic rays, and there is no reason to suppose that Anderson, Neddermeyer, Street or Stevenson knew about the problems of strong forces. The speculation and the experiment were quickly put together by people like Niels Bohr and it was supposed that Yukawa's theory applied to the mesons discovered by the experimenters.

We know exactly when and how the dubbing of the experimental particle took place. Millikan wrote to the *Physical Review* as follows:[3]

After reading Professor Bohr's address at the British Association last September in which he tentatively suggested the name 'yucon' for the newly discovered particle, I wrote to him incidently mentioning the fact that Anderson and Neddermeyer had suggested the name 'mesotron' (intermediate particle) as the most appropriate name. I have just received Bohr's reply to this letter in which he says:

'I take pleasure in telling you that every one at a small conference on cosmic-ray problems, including Auger, Blackett, Fermi, Heisenberg, and Rossi, which we have just held in Copenhagen, was in complete agreement with Anderson's proposal of the name "mesotron" for the penetrating cosmic-ray particles.'   Robert A. Millikan
California Institute of Technology
Pasadena, California
December 7, 1938

Note that Bohr had suggested the name 'yucon' in honour of Yukawa, but the experimentalist's name took hold by unanimous consent. Indeed there were problems from the start about the 1936 particle being what Yukawa needed – the calculated and actual lifetimes were utterly discrepant. Much later, in 1947, another particle was found in cosmic rays, while the new accelerators were starting to verify the existence of a range of related particles in scattering experiments. These were the kind of thing that Yukawa had wanted, and they came to be called π-mesons. The 1936 particle became a μ-meson. After a while it was evident that they were totally different kinds of thing – a π-meson and a μ-meson being

3 This letter was published in *The Physical Reviews* 55 (1939) p. 105. The papers using the Bethe–Heitler energy-loss formula to reveal the original mesons (muons) are S.H. Neddermeyer and C.D. Anderson, *ibid.* 51 (1937), pp. 884–6, relying on data and photographs in *ibid.* 50 (1936), pp. 263–7. Also J.C. Street and E.C. Stevenson, *ibid.* 51 (1937), pp. 1005A.

about as unlike as any pair of entities in nature can be. The name 'meson' stuck with the post-1947 particles, and the 1936 particle became a muon. Histories of the subject now imply that Anderson *et al.* were actually out looking for an object to fit Yukawa's conjecture – a conjecture they had never even heard of!

I shall return later to the question, Which comes first, theory or experiment? Chapter 9 has more examples of how theory-obsessed histories turn experimental explorations into investigations of a theory which was totally unknown to the experimenters.[4] For the present our concern is reference. The meson/muon story does not fit well with Putnam's meaning of 'meaning'. Putnam wanted to make the reference, in the end, the lynchpin of meaning. The name would apply to an entity that had been dubbed by that name on a particular historical occasion, at a baptism as it were. In our case, there was such a baptism, in 1938. However, the very name 'mesotron' or 'meson' came to mean, for the theoreticians, 'whatever it is that satisfies Yukawa's conjecture'. In short, the name acquired a sort of Fregeian sense. That is what took hold, baptism or no. When it was realized that this sense did not apply to the baptized object, the baptism was annulled, and a new dubbing took place.

Meaning

Putnam's theory of meaning works well for success stories like electrons. It is imperfect around the edges. It leaves us unhappy about bifurcating concepts, such as acidity. It does not explain how people with different theories about a nonentity such as caloric can communicate just as well with each other as people with different theories about real entities, say electrons. It relies in part on historical dubbings, the benefit of the dubbed, and a causal chain of the right sort passing from the first baptism to our present use of a name. Real communities cheerfully disregard baptisms if they want to. Those who wish a theory of meaning for scientific terms will have to improve on Putnam. They will also pay attention to the contrast between Putnam's story and what happens, in real life, in the life sciences. This contrast has been well described by John

---

4 In a letter to C.W.F. Everitt about our joint project, 'Theory or experiment, which comes first?', the Nobel laureate physicist E. Purcell suggested numerous examples of the way theory rewrites experimental history. Checking his case of the μ-meson led me to use the example to illustrate reference-stealing as above.

Dupré.[5] I have only one admonition. When philosophers turn to this topic, let them not wave their hands, henceforth, about dubbings and baptisms and so forth. Let them, like Dupré, look for example at taxonomy. Let us not speak about dubbing in the abstract, but about those events in which glyptodons, caloric, electrons or mesons were named. There is a true story to be told of each. There is a real letter written by Millikan. There is a real getting together of Frenchmen to name substances, including caloric. There was even a real Johnstone Stoney. The truths about those events beat philosophical fiction any day.

I have not wanted to advance a philosophical theory of meaning. I have had only the negative purpose, of describing a theory of meaning that is pretty natural for a wide range of linguistic practice, and which does not invite talk of incommensurability. It is the kind of theory that scientific realists about entities need. It is especially attractive if one is rather cool about realism about theories. For if one expects that our theories are not strictly true, one will not want to use them to define entities in any permanent way. Rather one wants a notion of reference that is not tied by any specific, binding theory about what is referred to. A Putnamian account of reference does not, however, force you to be a realist. We must now consider why Putnam has abandoned out-and-out realism.

5 'Natural kinds and biological taxa', *The Philosophical Review* 90 (1981), pp. 66–90.

7   Internal realism

This chapter is probably irrelevant to scientific realism and so can well be omitted. It is about Putnam's important new 'internal realism', apparently a species of idealism.[1] A switch from realism to idealism sounds central to our discussion, but it is not. Putnam is no longer engaged in the debate between the scientific realist and the anti-realist about science. That debate makes a keen distinction between theoretical and observable entities. Everything Putnam now says ignores that. So it should be. His is a philosophy founded upon reflections on language, and no such philosophy can teach anything positive about natural science.

To omit Putnam's developments would nevertheless be to bypass issues of current interest. Moreover, since he finds a predecessor in Kant, we can bring in Kant's own kind of realism and idealism. Kant is a useful foil to Putnam. If we simplify and pretend that Kant too is an 'internal realist' (or that Putnam is a 'transcendental idealist') we can imagine a Kant who, unlike Putnam, emphasizes the difference between observed and inferred entities. Putnam seems to be a scientific realist within his internal realism, while we can invent a Kant who is an anti-realist about theoretical entities within a similar setting.

Internal and external realism

Putnam distinguishes two philosophical points of view. One is 'metaphysical realism', with an 'externalist perspective' about entities and truth: 'the world consists of some fixed totality of mind-independent objects. There is exactly one true and complete description of "the way the world is". Truth involves some sort of correspondence between words or thought-signs and external things and sets of things' (p. 49).

1 All references to Hilary Putnam in this chapter are to his *Reason, Truth and History*, Cambridge, 1982.

Putnam proposes instead an 'internalist perspective' which holds that the question,

> *what objects does the world consist of?* is a question that it only makes sense to ask *within* a theory or description. . . . 'Truth', in an internalist view, is some sort of (idealized) rational acceptability – some sort of ideal coherence of our beliefs with each other and with our experiences *as those experiences are represented in our belief system.*

At this level internalism and pragmatism have much in common. Putnam's position depends additionally on ideas about reference. He rejects metaphysical realism because there is never any hook-up, or correspondence, between my words and a particular batch of mind-independent entities. 'Objects' do not exist independently of conceptual schemes. 'We cut up the world into objects when we introduce one sign or another. Since the objects *and* the signs are alike *internal* to the scheme of description, it is possible to say what matches what.' (p. 52).

Putnam reports another difference between metaphysical and internal realism. The internalist says truth is optimal adequacy of theory. The externalist says that truth is, well, truth.

Internalist: If we had a complete theory of everything in the universe of interest to us, and the theory was thoroughly adequate by current standards of warranted assertability, rationality, or whatever, then that theory would, by definition, be true.

Externalist: Such a theory would very probably be true. But it is conceivable that the adequacy is a matter of luck or demonology. The theory might work for us, and yet still be a false theory about the universe.

Queries about metaphysical realism

Putnam's internalist can make no sense of a complete theory of the interesting universe which is entirely adequate but still false. I'm an externalist, and can make no sense of it either, *but for a different reason.* I cannot understand the idea of a complete theory of our interesting universe. *A fortiori*, I don't understand the idea that such a theory might be adequate but false, for the idea of such a theory is itself incoherent. I can contemplate a complete theory for those wretched so-called possible worlds envisaged by logicians, but for our world? Balderdash.

Four articles were advertised on a flysheet for the April 1979 *Scientific American*: How the bare hand strikes a karate blow; An enzyme clock; The evolution of disc galaxies; Oracle bones of the Shang and Chow dynasties. How could there be a complete theory of even those four topics, let alone a complete and unified theory of *everything* (including these four topics)?

How indeed could there be a complete account of even one thing or one person? P.F. Strawson remarks in his book *Individuals*, 'The idea of an "exhaustive description" is in fact quite meaningless in general' (p. 120). Strawson was then writing about Leibniz. Leibniz might be the best candidate for a metaphysical realist. He did think that there is a body of truth external to our own beliefs. He probably did think that there is one best, divine, description of the universe. He did think there is one totality of basic objects, namely monads. I don't suppose he thought they are 'mind-independent' since monads are minds, more or less. But Leibniz did not hold a correspondence theory of truth. Even Leibniz does not fill Putnam's bill. Was any serious thinker a metaphysical realist?

Maybe it does not matter. Putnam was describing a certain perspective, rather than a definitive theory of reality. We well recognize that externalist perspective. But here we must be careful. There could be some instances of that perspective – some kinds of external realism – which are immune to Putnam's objections, because his objections are directed at metaphysical realism as *he* defined it.

For example, take his phrase in the definition: 'fixed totality of mind-independent objects'. Why fixed? Why one totality? Consider only the banal example of Eddington's – there are two tables, namely the wooden table at which I am writing, and a certain bundle of atoms. A realist about entities can well hold (a) there are mind-independent tables, (b) there are mind-independent atoms, and (c) no set of atoms is identical with this table at this instant. Atoms and tables have to do with different ways of carving up the world. There is no one fixed totality of objects. A Rubik's cube may be a totality of 27 smaller cubes, but it need not be the case that each of these is a totality of atoms which taken together are the totality of the Rubik's cube.

Do I not then grant Putnam's assertion, quoted above? We cut up the world into objects when we introduce one or another scheme of description. Yes, I grant that, metaphorically speaking. I do not grant the preceding sentence, '"Objects" do not exist independently of conceptual schemes.' There are both atoms and Rubik's cubes. To take another trite example, Inuit are said to distinguish ever so many kinds of snow that look pretty much the same to us. They cut up the frozen North by introducing a scheme of description. It in no way follows that there are not 22 distinct mind-independent kinds of snow, precisely those distinguished by the Inuit. For all I know, the powder snow, corn snow, or Sierra cement spoken of by some skiers neither contain nor are contained in any Inuit class of snow. The Inuit do not ski, and may never have wanted that category. I expect that there is still powder snow *and* all the Inuit kinds of snow, all real mind-independent distinctions in a real world.

These remarks do not prove that there is powder snow, whether anyone thinks of it or not. They merely observe that the fact that we cut up the world into various possibly incommensurable categories does not in itself imply that all such categories are mind-dependent.

Let us then be wary of the way in which Putnam runs a number of different theses together as if there were some logical connection between them.

Metaphysical fieldwork

Putnam, I said, was a scientific realist who has become something of an anti-realist. Did he change sides? No. To use a gruesome analogy, he changed wars.

Scientific realism, opposed to anti-realism about science, is a *colonial* war. The scientific realist says that mesons and muons are just as much 'ours' as monkeys and meatballs. All of those things exist. We know it. We know some truths about each kind of thing and can find out more. The anti-realist disagrees. In the positivist tradition from Comte to van Fraassen, the phenomenal behaviour of meatballs and monkeys may be known, but talk about muons is at most an intellectual construct for prediction and control. Anti-realists about muons are realists about meatballs. I call this a colonial war because one side is trying to colonize new realms and call them reality, while the other side opposes such fanciful imperialism.

Then there is *civil* war, between say Locke and Berkeley. The realist (Locke) says that many familiar entities have an existence independent of any mental goings on: there would be monkeys even if there were no human thoughts. The idealist (Berkeley) says everything is mental. I call this a civil war because it is fought on the familiar ground of everyday experience.

Civil wars need not be fought only on home territory. Berkeley fought a colonial war too. He detested the corpuscular and mechanical philosophy of Robert Boyle. It said, in the extreme, that matter consists of bouncy springlike corpuscles (molecules, atoms, and particles, as we would say). Berkeley fought a colonial war partly because he thought that if he won, the imperialist home government of realism/materialism would collapse. Matter would be vanquished by mind.

Finally there is *total* war, chiefly a product of more recent times. Maybe Kant began it. He rejects the assumptions of civil war. Material events occur with as much certainty as mental ones. There is indeed a difference between them. Material events happen in space and time, and are 'outer', while mental events happen in time but not in space, and are 'inner'. But I can know that the meatball on my plate is mush exactly as well as I know that my emotions are confusing. In general I no more infer the mushiness from my sense data than I infer that I am mixed up from my behaviour (though I could do either, on occasion).

Putnam once argued for scientific realism in a colonial war. He now argues for a position, which he says is like Kant's, in a total war. Let us grasp Kant's position in more detail before tackling Putnam's.

Kant

Kant watched his predecessors engaged in civil war. On one side there was Locke's thesis. Kant calls it *transcendental realism*: there are objects really out there, and we infer their existence and their properties from our sense experience. Then there was Berkeley's antithesis. Kant calls it *empirical idealism*. Matter itself does not exist; all that exists is mental.

Kant invented a synthesis to turn all this upside down. He literally reverses the labels. He calls himself an *empirical realist* and a *transcendental idealist*.

He did not go directly to his final position, but approached it by another duality. Is space merely a relative notion, as Leibniz urged and Einstein is supposed to have established? Or is it absolute, as in the Newtonian scheme? Newton had a thesis, that space and time are real. Objects occupy positions in a predetermined space and time. Leibniz voiced an antithesis, that space and time are not real. They are ideal, that is, constructs out of the relational properties of objects. Kant shilly-shallied between the two for most of his life, and then created a synthesis. Space and time are preconditions for the perception of something as an object. It is not an empirical fact that objects exist in space and time although we may experimentally determine the spatio-temporal relationships of objects within the framework of space and time. This is an *empirical realism* that grants 'the objective validity, of space, in respect of whatever can be presented to us outwardly as object'. At the same time it is a *transcendental idealism* which asserts that space 'is nothing at all . . . once we withdraw . . . its limitation to possible experience and so look upon it as something that underlies things in themselves' (p. 72).[2] It took Kant another decade to make this approach fit the whole range of philosophically problematic concepts. Berkeley the immaterialist had denied the existence of matter and the externality of external objects. There is nothing but mind and mental events. Kant's response: 'Matter is . . . only a species of representations (intuition) which are called external, not as standing in relation to objects in themselves external, but because they relate perceptions to the space in which all things are external to one another, while yet the space itself is within us'. Thus space itself is ideal, 'within us', and matter is properly called external because it exists as part of a system of representation within this ideal space. In order to arrive at the reality of outer objects I have just as little need to resort to inference as I have in regard to the reality of the objects of my inner sense, that is, in regard to the reality of my thoughts. For in both cases alike the objects are nothing but representations, the immediate perception (consciousness) of which is at the same time a sufficient proof of their reality.

The transcendental idealist is, therefore, an empirical realist. It is essential to Kant's point of view that what we call objects are

2 All quotations from Kant are from the N. Kemp Smith translation of *The Critique of Pure Reason*, London, 1923.

constituted within a scheme, and that all our knowledge can pertain only to objects thus constituted. Our knowledge is of phenomena, and our objects lie in a phenomenal world. There are also noumena, or things in themselves, but we can have no knowledge of these. Our concepts and categories do not even apply to things-in-themselves. Philosophers from Hegel on have usually dismissed Kant's things-in-themselves. Putnam, warming to Kant, expresses gentle sympathy for the idea.

Truth

According to Putnam, 'although Kant never quite says that this is what he is doing, Kant is best read as proposing for the first time what I have called the "internalist" or "internal realist" view of truth' (p. 60). Like so many modern philosophers, Putnam builds much of his philosophy around the idea of truth. Of Kant he says that 'there is no correspondence theory of truth in his philosophy'. Not surprising: There is *no* theory of truth in Kant's philosophy! Kant's concerns were not Putnam's. So far as affects realisms, he had two main problems:

Are space and time real or ideal, Newtonian or Leibnizian?

Are external objects mind-independent and Lockeian, or is everything mental and Berkeleyan?

His empirical realism and transcendental idealism is a synthesis of these oppositions and has little to do with truth. Yet Putnam's injection of a theory of truth into Kant is not strictly wrong. Putnam attributes to Kant the following ideas:

Kant does not believe that we have objective knowledge.
The use of the term 'knowledge' and 'objective' amount to the assertion that *there is still a notion of truth.*
A piece of knowledge (i.e. a 'true statement') is a statement that a rational being would accept on sufficient experience of the kind that it is possible for beings with our nature to have.
Truth is ultimate goodness of fit. (p. 64).

Perhaps Putnam hit the nail on the head, particularly since he himself tends towards the pragmatist idea that truth is whatever a rational community would in due course find coherent and agree to. Kant wrote:

The holding of a thing to be true is an occurrence in our understanding which, though it may rest on objective grounds, also requires subjective causes in the mind of the individual who makes the judgement. If the judgement is valid for everyone, provided only that he is in possession of reason, its ground is objectively sufficient. . . . Truth depends upon agreement with the object, and in respect of it the judgements of each and every understanding must therefore be in agreement with each other. . . . The touchstone, whereby we decide whether our holding a thing to be true is objective, is the possibility of communicating it and of finding it to be valid for all human reason. For there is then at least a presumption that the ground of the agreement of all judgements with each other, notwithstanding the differing characters of individuals, rests upon the common ground, namely, upon the object, and that it is for this reason that they are all in agreement with the object – the truth of the judgement being thereby proved (p. 645).

To what extent does this make Putnam mesh with Kant? I leave that to the reader. Putnam thinks warranted rational assertability and truth go hand in hand. Kant also wrote, ' I cannot *assert* anything, that is, declare it to be a judgement necessarily valid for everyone, save as it gives rise to' universal agreement among reasoning people (p. 646).

Theoretical entities and things-in-themselves

Scholars do not agree about Kant's noumenal world of things in themselves. Putnam reads Kant as saying that not only can we not describe things-in-themselves, but also 'there is not even a one-to-one correspondence between things-for-us and things-in-themselves'. There is no horse-in-itself corresponding to the horse in the field. There is only the *noumenal world* which as a whole somehow 'gives rise to' our system of representation.

There have been quite different traditions of interpretation. One holds that theoretical entities are Kant's things-in-themselves. I first find this in J.-M. Ampère (1775–1836), founder of the theory of electromagnetism. Deeply influenced by Kant, he could not tolerate the anti-realist impulses set loose on the world. He insisted that we can postulate noumena, and laws between them, to be tested in experience. This postulational and hypothetico-deductive method, said Ampère, is an intelligent investigation of the noumenal world. In our day the philosopher Wilfred Sellars holds a similar view.

There may even be an important connection, in the development of Kant's own thought, between noumena and theoretical entities. In 1755, when he was young, Kant wrote a small physics tract called *Monadology*. This is a remarkable anticipation of our modern theory of fields and forces. Two years later Boscovič elaborated it with far greater mathematical skill and launched field theory on the world. In Kant's early physics the world is made up of point particles – monads – separated by finite distances and exercising force fields in their environment. The properties of matter were explained by the resulting mathematical structure. *In 1755 these theoretical point particles of Kant's were his noumena.* Much later he revised this idea, and realized that there was a formal inconsistency in his theories. It could be resolved only by eliminating the things, the point particles, leaving nothing but fields of force. As a result, in the underlying structure of the universe, *there are no things, no noumena.* Then came the usual Kantian synthesis of these conflicting propositions: there are no *knowable* noumena.

Thus it is tempting to suggest that Kant's doctrine about things in themselves arose as much from his physics as his metaphysics. Kant was of little value as a scientist, but he would have been a wonderful member of a panel for a national science foundation, disbursing research money among widely different projects. He picked winners. There is what we now call the Kant–Laplace hypothesis, about the formation of the solar system. He was from the start on the side of evolutionary hypotheses about the origins of the species and the human race. He picked field theories over atomistic approaches. Now, the state of knowledge appropriate to his century was one which would downplay the significance of theoretical entities as things in themselves. There were, indeed, hypothetical stuffs of various sorts, such as the electric fluids of Franklin and many others, or the magnetic poles of Coulomb. There was an immense amount of talk about Newtonian particles and forces, but it was only at the time of Kant's death, just after the beginning of the nineteenth century, that these really got going again. Kant's attitude to the thing in itself is a quasi-scientific reaction to the modifications in his 1755 programme. Ampère, the first to preach that after all there are knowable noumena, namely the theoretical entities of the new physics and chemistry, reflects the transformation in physics. He began his career as a chemist, and was preaching knowable noumena almost as soon as he had mastered the new conjectures about the atomic structure of the elements.

What position ought Kant to have taken about theoretical entities that really do some work in science? What would he have done when, in the twentieth century, we learned how to manipulate and even spray electrons and positrons? His own realism/idealism was directed at familiar observable objects. He denied that we infer them for our sense-data. Theoretical entities are in contrast inferred from data. Would Kant have been an empirical realist about chairs, that need no inferring, while staying an empirical anti-realist about electrons? That seems to be a possible position.

Reference

Putnam's most original contribution concerns reference more than truth. His meaning of 'meaning' described in the previous chapter contains the seeds of its own decay. They are plain to see, for they are none other than what I called the 'dots of extension'. The meaning of a natural kind term is a sequence of elements ending in the extension, but you can't write that down.

Putnam first thought that unlike Fregeian senses, reference was unproblematic. The reference of 'glyptodon' could be indicated by pointing at a skeleton and some features in the stereotype. If glyptodons form a natural kind, nature would do the rest, and determine the extension. Theoretical entities could not be pointed at, but were to be handled by an historical story about the introduction of the terms that denote them, plus some charitable principles of the benefit of the doubt.

Putnam became sceptical. The malaise about meanings and Fregeian senses owes much to W.V.O. Quine's doctrine of the indeterminacy of translation. Quine had a parallel thesis about reference: the inscrutability of reference. To put the idea crudely: you can never tell what someone else is talking about, nor does it matter much. Quine asserted this with modest examples: where I speak of rabbits you might hear me as talking about spatio-temporal slices of rabbithood. Putnam adds real inscrutability. Whenever you talk of cats and mats, you might be referring to what I refer to when I speak of cherries and trees – yet the difference in reference would not come out, because anything I am confident of (some cat is on some mat) is expressed by a sentence which under your interpretation is something in which you have equal confidence (some cherry is on some tree).

This is indeed extraordinary. We are under two difficulties. We need to have this bizarre claim made plausible to us, and we need to understand its place in the argument against external or meta-physical realism. Thus we need to have a local argument for the cat/cherry conclusion, and we need to have a global argument, showing how that leads to anti-metaphysical position.

Cats and cherries

No view which only fixes the truth-values of whole sentences can fix reference, even if it specifies truth values for sentences *in every possible world.*

That is Putnam's theorem (p. 33), which we shall explain. Its cash value is presented in terms of cats and cherries. Every time you speak of cherries, you could be referring to what I call cats, and vice versa. Were I seriously to say that a cat is on a mat, you would assent, because you took me to be saying a cherry is on a tree. We can reach total agreement on the facts of the world – that is, on the sentences we hold to be true – and yet it might never appear that when I am talking about cats, you are talking about what I call cherries. Moreover your system of reference could systematically so differ from mine that the difference between us could not come out, no matter what is true about cats and cherries.

This striking conclusion follows a well-known result in mathematical logic, called the Löwenheim–Skolem theorem. The basic idea is the result of work by L. Löwenheim in 1915 and developed by Th. Skolem in 1920. In that era it seemed plausible to try to characterize mathematical objects, such as sets, by means of postulated axioms. An intended object, such as a set, would be something that fitted some postulates, and so the postulates would define the class of intended objects. Moreover we hoped to do this in the only well-understood branch of logic, called first-order logic – the logic of the sentential connectives ('and', 'not', 'or', or whatnot) and first-order quantifiers ('all', 'some').

It was thought by logicians of the day that some kind of theory of sets could serve as the foundations for many or all branches of mathematics. Georg Cantor proved a famous result. He first had clarified the idea of some infinite sets being bigger than others. Then he showed that the set of subsets of natural numbers is bigger than the set of natural numbers. In another formulation, he showed that the set of all real numbers, or of all numbers expressed as decimal numbers, is larger than the set of natural numbers. Once this fact had been digested and accepted by classical logicians, Löwenheim and Skolem proved something that at first seemed paradoxical.

You write down some postulates that you hope capture the very essence of sets built up from sets of natural numbers. Within these postulates you prove Cantor's theorem, which says that the set of subsets of natural numbers is not denumerable, that is, it cannot be paired off with the natural numbers and so is bigger than the set of natural numbers itself. So far so good. In the way in which you intend your postulates to be understood, you are talking of Cantorian sets. Löwenheim and Skolem proved, however, that any theory, expressed in first-order logic, which is true of some domain of objects, is also true of a denumerable domain. Thus you intended your postulates to be true of Cantorian sets. Cantor's theorem at once convinces us that there are more Cantorian sets than there are natural numbers. But those very same postulates can be re-interpreted so as to be true of a much smaller domain. Suppose P is the sign which, in your theory, denotes the set of all subsets of the set of natural numbers. That is bigger than the set of natural numbers. Your theory can be reinterpreted so that P denotes something surely different, a set no bigger than the set of natural numbers.

The Löwenheim–Skolem theorem once seemed paradoxical, but it has now been digested. Most students of logic find it rather obvious, natural, and inevitable. They say things like, 'in a first-order formulation, there have to be nonstandard models'.

Putnam returns the theorem to seeming paradox. He makes a correct generalization. It applies to any domain of individuals, say cats and cherries. Take as the axioms all truths about these – all truths that I shall ever utter, or that people will ever utter, or simply all the genuine truths expressible in the first-order language. Whatever you choose, there will be unintended interpretations: moreover when we pick two kinds of objects, cats and cherries, and use a short list of truths, we can get the intended interpretation about cats to map on to the unintended interpretation about cherries. Putnam provides the details both for the short example and for the full theorem.

The implications for scientific realism

Putnam supposes that these technical results are bad for scientific realism. Why? Largely because he thinks that scientific realism is in the end a copy or correspondence theory of truth. Our theories are true because they represent the world, and they latch on to the world by referring to objects – a reference which Putnam now thinks makes sense only within a system of beliefs.

Much of this position is well known. It is a longstanding criticism of correspondence theories that the sentences are supposed to correspond to facts, but there is no way to distinguish the facts except in terms of the sentences to which they correspond. G.E. Moore is not notable for his anti-realism, but here is how he expressed the idea 80 years ago, in an article on 'Truth' published in Baldwin's *Dictionary of Philosophy*:

It is commonly supposed that the truth of a proposition consists in some relation which it bears to reality; and falsehood in the absence of this relation. The relation in question is generally called a 'correspondence' or 'agreement', and it seems to be generally conceived as one of partial similarity; but it is to be noted that only propositions can be said to be true in virtue of their partial similarity to something else, and hence that it is essential to the theory that a truth should differ in some specific way from the reality, in relation to which its truth is to consist, in every case except that in which the reality is itself a proposition. It is the impossibility of finding any such difference between a truth and the reality to which it is supposed to correspond which refutes the theory.

It has been argued, for example by J.L. Austin, that correspondence theories do have merit, because, contrary to Moore, there is an independent way to pick out facts. There are, first of all, independent ways to pick out things and qualities we are talking about – by pointing, for example. Then we make assertions by connecting referring expressions and names for properties and relations. A proposition is true just if the property named is possessed by the object referred to. Putnam must suppose that his use of the Löwenheim–Skolem vitiates this Austinian move, by showing once again that there is no way to make independent reference. But all he has shown is that you cannot succeed in reference by stating a set of truths expressed in first-order logic. When we look more closely at the Löwenheim–Skolem theorem, we recall that it has premises. There are ways of evading these premises and thus casting doubt on Putnam's conclusions.

Premises

1 The Löwenheim–Skolem theorem is about sentences in first-order logic. No one has ever shown that the commonplace language of physicists can ever be squeezed into a first-order format. So the argument is not known to be relevant to, say, quantum electro-dynamics, and hence not to scientific realism.

2 There is a weighty school of thought, deriving impetus from the late Richard Montague, that ordinary English primarily deploys second-order quantifiers. In no direct way does the Löwenheim–Skolem theorem extend to such languages, so the applicability of Putnam's work to plain prescientific English is controversial.

3 Much common speech involves what are called indexicals. These are words whose reference depends on the context of utterance: this, that, you, me, here, now (not to mention our tensed verbs). As I walk out this fine morning I overhear: 'Hey you, stop picking my cherries, come here this instant!' Only dogma could insist that this ordinary sentence is expressible in first-order logic.

4 Introduction of indexicals goes only part of the way. Indexicals are pointers, but they are still linguistic. Language is embedded in a wide range of doings in the world. Putnam oddly refers to Wittgenstein during his discussion, recalling Wittgenstein's argument that meanings cannot be exhaustively given by rules. That did not mean, for Wittgenstein, that there was something intrinsically indeterminate and open to reinterpretation in our linguistic *practice*. It meant that language is more than talking. This is no place to expound a version of his insights, but cherries are for eating, cats, perhaps, for stroking. Once speech becomes embedded in action, talk of Löwenheim and Skolem seems scholastic. They were entirely right in what they said about a certain view of mathematical objects. They wisely refrained from discussing cats. We can do nothing with very large numbers except talk about them. With cats we relate in other ways than speech.

5 Putnam says that whatever theory we propound about reference and denotation, words such as 'denote' and 'refer' can themselves be reinterpreted. Suppose I say that 'cat' denotes

animals like those on my lap. He asks: How do I know that 'denotes' denotes denoting? But of course I never normally use words such as 'denote' in explaining the usage of words. That function may be served by 'That is a glyptodon skeleton', used to explain what a glyptodon is. I do not need a theory of reference in order to refer, and it is at least arguable, on grounds possibly learned from Wittgenstein, that there could be no general theory of reference.

6 Putnam is writing about unscientific anti-realism, so it is right to discuss cherries and cats. Might we not grant him his point for the theoretical entities of natural science? Is not the dubbing of entities with names entirely at the level of language? No, often it is not. Look at the 1936 paper of Anderson and Neddermeyer, mentioned in the last chapter. That is the one with the data on the basis of which the physics community dubbed the mesotron or meson – later muon. The paper is full of photographs. Not snapshots of muons, but tracks. It measures angles between the tracks caused by the collisions of this and that. We do use indexicals as brief as 'this' and 'that' to point to the most theoretical of entities – not by pointing at them, but by pointing at their traces. Not that we stop there. As is clear from my previous chapter, people at first were pretty unsure about those things that came to be called muons. But now for example we know that the mass of the muon is 206.768 times that of the electron.

This last sentence will seem grist to Putnam's mill. For that is just the sort of truth we can put in as an axiom in an account of muons. Can we not then expose it to Löwenheim–Skolem reinterpretation? I do not think so, for how did we get this fine number to three places of decimals? It is a rather complicated computation in which we determine a whole bunch of quantities, such as the magnetic moment of the free electron, the Bohr magneton, and other fancy stuff, and in particular, relationships between a number of constants of nature. Now if these were just a bunch of sentences, and we could do all the mathematical physics in terms of first-order logic, the Löwenheim–Skolem theorem would apply. But in every case the numbers and ratios are intimately connected to specific experimental determinations. These in turn are all connected up with people, places, and, above all, doings. (Typical example: the University of Washington–Lawrence Radiation Laboratory group, i.e. K.M. Crowe, J.F. Hague, J.E. Rotherberg, A. Schenck, D.L.

Williams, R.W. Williams and K.K. Young, *Phys. Rev. D.* 2145 (1972).) Nor is it just one such set of doings, but lots of independent but not totally dissimilar *doings* all over the world.

7 Putnam does address the question of whether humans could ever use his unintended interpretation of the word 'cat'. He notes a symmetry between intended and unintended interpretations – everything we explain in terms of cats, others can explain in terms of cherries. He reiterates a discussion that derives from Nelson Goodman's book *Fact, Fiction and Forecast*. There is an important fact that he ignores. The Löwenheim–Skolem theorem is non-constructive. That is, there is in principle no humanly accessible way to generate an unintended interpretation.

8 Nor do we need technical examples to begin to query Putnam's confidence. Putnam cites his colleague Robert Nozick, as suggesting that (in Putnam's view) all women might mean cats when they speak of cherries, whereas 'we' men mean cherries. But there are for example nominal adjectives, illustrated by Bing cherries and Persian cats. Nominal adjectives such as 'Bing' are not ordinary modifiers like 'sweet', for sweet Bing cherries are sweet fruit, but they are not 'Bing fruit'. How is the Putnam/Nozick reinterpretation continued? Do their fantasy women mean Persian cats when they speak of Queen Anne cherries? That is, does one kind of cherry map on to one kind of cat? That won't do, for the number of kinds of cherries is different from the number of kinds of cats, so no such mapping will preserve the structure of nominal adjectives. More importantly, Queen Anne cherries are for preserving or for pies, while Bing cherries are for eating ripe from the tree. How are these facts to show up in the structure of facts about cats?

Putnam perhaps commits one of the gravest errors of philosophy. He has an abstract theorem. Then he explains its content in terms of one sentence that no one before him has ever uttered, nor would commonly have any point outside logic in uttering: 'Some cherry is on some tree'. Then he passes to the assertion that just as you can reinterpret 'cherry' you can reinterpret 'denote'. All the flourishing ordinary world of making a pie out of Queen Anne cherries, of determining the ratio of the masses of muons and electrons – all that is left out.

I shall not continue. I wanted only to emphasize that (a) assuring reference is not primarily a matter of uttering truths, but of interacting with the world, and that (b) even at the level of language there is vastly more structure than Putnam discusses, be it deep questions about the language of mathematical physics, or trivial observations about Bing cherries.

Nominalism

The above reflections do not mean you need dissent from Putnam's underlying philosophy. They mean only that what looks like a spiffy argument needs more polishing than it has yet received. What is the underlying point of view? I have followed Putnam in comparing his ideas to Kant, but there is a significant difference. Kant called himself a transcendental idealist. I would call Putnam a transcendental nominalist. Both are kinds of anti-realism. Before Kant, realism usually meant anti-nominalism. After Kant, it usually meant anti-idealism.

Idealism is a thesis about *existence*. In its extreme form it says that all that exists is mental, a production of the human spirit.

Nominalism is about *classification*. It says that only our modes of thinking make us sort grass from straw, flesh from foliage. The world does not have to be sorted that way; it does not come wrapped up in 'natural kinds'. In contrast the Aristotelian realist (the anti-nominalist) says that the world just comes in certain kinds. That is nature's way, not man's.

The idealist need have no opinion about classification. He may hold that there is indeed a real distinction between grass and straw. He says only that there is no stuff, grass and straw; there are only ideas, mental entities. But the ideas could well have real essences.

Conversely the nominalist does not deny that there is real stuff, existing independent of the mind. He denies only that it is naturally and intrinsically sorted in any particular way, independent of how we think about it.

In fact nominalism and idealism tend to be part of the same cast of mind. That is one reason that the word 'realism' has been used to denote opposition to either doctrine. But the two are logically distinct.

I read Kant in a possibly extreme way. He thought that space and time are ideal. They literally do not exist. Although there are empirical relations determinable within space and time, those relations, being spatio-temporal, have no existence beyond the mind. Kant was indeed a transcendental *idealist*. Putnam is instead a transcendental *nominalist*.

Putnam's internal realism comes to this: Within my system of thought I refer to various objects, and say things about those objects, some true, some false. However, I can never get outside my system of thought, and maintain some basis for reference which is not part of my own system of classification and naming. That is precisely empirical realism and transcendental nominalism.

Revolutionary nominalism

T.S. Kuhn has also been read as an idealist. I think he too is better understood as a transcendental nominalist – one who got there before Putnam. But whereas Putnam's reflections are based on an *a priori* theorem and alleged implications for language, Kuhn has more of a real-life basis for his position.

A scientific revolution, in Kuhn's opinion, produces a new way of addressing some aspect of nature. It provides models, conjectured laws, classes of entities, causal powers which did not enter into the predecessor science. In a completely uncontroversial sense we may now live in a different world from the nineteenth-century age of steam – a world in which aeroplanes are everywhere and railways are going bankrupt. More philosophically (perhaps) it is a different world, in that it is categorized in new ways, thought of as filled with new potentialities, new causes, new effects. But this novelty is not the production of new entities in the mind. It is the imposition of a new system of categories upon phenomena, including newly created phenomena. That is why I call it a kind of nominalism. Here is a recent formulation of Kuhn's own:

What characterizes revolutions is, thus, change in several of the taxonomic categories prerequisite to scientific descriptions and generalizations. That change, furthermore, is an adjustment not only of criteria relevant to categorization, but also of the way in which given objects and situations are distributed among pre-existing categories. Since such redistribution always involves more than one category and since those categories are interdefined, this sort of alteration is necessarily holistic.[3]

Kuhn is no old fashioned nominalist. That would be someone who thought that all our classifications were a product of the human

---

3 T.S. Kuhn, 'What are scientific revolutions?' Center for Cognitive Science Occasional Paper 18, Massachusetts Institute for Technology, 1981, p. 25.

mind, not the world, and that those classifications were all the same absolutely stable features of our minds. He can disagree with that nominalist on both counts. Obviously he favours the possibility of revolutionary change, and he furnishes us with examples of it. He can equally assert that many of our prescientific categories *are* natural kinds: people and grass, flesh and horseflesh. The world simply does have horses and grass in it, no matter what we think, and any conceptual scheme will acknowledge that. There is no reason that the history of science should deny that the world sorts itself in these ways. Nor is there much reason, in the comparative study of cultures, to suppose that any other people fail to sort in similar ways. Kuhn's nominalism, in so far as it is founded upon his historical studies, could teach only that some of our scientific categories can be dislodged. Time-honoured categories, such as substance and force, may go under. Time and space may even take a beating. Kuhn does teach a certain relativism, that there is no uniquely right categorization of any aspect of nature. Indeed the idea of an aspect of nature, of comprising just such and such affairs, is itself a variable. The Greeks, we say, had no concept of electricity, Franklin no concept of electricity-and-magnetism. Even such 'aspects of nature' emerge, weave in and out, during our history. The revolutionary nominalist infers that we have not reached the end of the road. Nor is the notion of an end of the road, of a final science, a truly comprehensible one.

The old-fashioned nominalist of times gone by held that our systems of classification are products of the human mind. But he did not suppose that they could be radically altered. Kuhn has changed all that. The categories have been altered and may be altered again. We can hardly avoid approaching nature with our present categories, problems, systems of analysis, methods of technology and of learning. We are in fact empirical realists: we think as if we are using natural kinds, real principles of sorting. Yet in the course of historical reflection we realize that the inquiries most dear to us may be replaced.

To sum up the idea: we do investigate nature as sorted into the natural kinds delivered by our present sciences, but at the same time hold that these very schemes constitute only an historical event. Moreover, there is no concept of *the* right, final representation of the world.

Putnam's remarks might incline one in the same direction, but there is one sense in which his present rendition is rather Kantian. Putnam has become conservative. For Kant there was no way out of our conceptual scheme. Putnam gives no reason to suppose there is any way either. Kuhn details ways in which there have been profound alterations. Thus his is a revolutionary transcendental nomenalism, whereas Putnam's is more conservative.

Rationality

There is another stand in Putnam's present position, reminiscent of Peirce. He holds that what is true is whatever we come to agree on by rational means, and he acknowledges that there may be at least evolution as we develop more and more styles of reasoning. I find it natural to explain this not in terms of Putnam's philosophy, but rather in terms of that of Imre Lakatos.

8   A surrogate for truth

'Mob psychology' – that is how Imre Lakatos (1922–74) carica-tured Kuhn's account of science. 'Scientific method (or "logic of discovery"), conceived as the discipline of rational appraisal of scientific theories – and of criteria of *progress* – vanishes. We may of course still try to explain changes in "paradigms" in terms of social psychology. This is . . . Kuhn's way' (I, p. 31).[1] Lakatos utterly opposed what he claimed to be Kuhn's reduction of the philosophy of science to sociology. He thought that it left no place for the sacrosanct scientific values of truth, objectivity, rationality and reason.

Although this is a travesty of Kuhn the resulting ideas are important. The two current issues of philosophy of science are epistemological (rationality) and metaphysical (truth and reality). Lakatos *seems* to be talking about the former. Indeed he is universally held to present a new theory of method and reason, and he is admired by some and criticized by others on that score. If that is what Lakatos is up to, his theory of rationality is bizarre. It does not help us at all in deciding what it is reasonable to believe or do now. It is entirely backward-looking. It can tell us what decisions in past science were rational, but cannot help us with the future. In so far as Lakatos's essays bear on the future they are a bustling blend of platitudes and prejudices. Yet the essays remain compelling. Hence I urge that they are about something other than method and rationality. He is important precisely because he is addressing, not an epistemological issue, but a metaphysical one. He is concerned with truth or its absence. He thought science is our model of objectivity. We might try to explain that, by holding that a scientific proposition must say how things are. It must correspond to the truth. That is what makes science objective. Lakatos, educated in Hungary in an Hegelian and Marxist tradition, took for granted the

---

1 All references to Imre Lakatos in this chapter are to his *Philosophical Papers*, 2 Volumes (J. Worrall and G. Currie, eds.), Cambridge, 1978.

post-Kantian, Hegelian, demolition of correspondence theories. He was thus like Peirce, also formed in an Hegelian matrix, and who, with other pragmatists, had no use for what William James called the copy theory of truth.

At the beginning of the twentieth century philosophers in England and then in America denounced Hegel and revived correspondence theories of truth and referential accounts of meaning. These are still central topics of Anglophone philosophy. Hilary Putnam is instructive here. In *Reason, Truth and History* he makes his own attempt to terminate correspondence theories. Putnam sees himself as entirely radical, and writes 'what we have here is the demise of a theory that lasted for over two thousand years' (p. 74). Lakatos and Peirce thought the death in the family occurred about two hundred years earlier. Yet both men wanted an account of the objective values of Western science. So they tried to find a substitute for truth. In the Hegelian tradition, they said it lies in process, in the nature of the growth of knowledge itself.

A history of methodologies

Lakatos presented his philosophy of science as the upshot of an historical sequence of philosophies. This sequence will include the familiar facts about Popper, Carnap, Kuhn, about revolution and rationality, that I have already described in the Introduction. But it is broader in scope and far more stylized. I shall now run through this story. A good many of its peripheral assertions were fashionable among philosophers of science in 1965. These are simplistic opinions such as: there is no distinction in principle between statement of theory and reports of observation; there are no crucial experiments, for only with hindsight do we call an experiment crucial; you can always go on inventing plausible auxiliary hypotheses that will preserve a theory; it is never sensible to abandon a theory without a better theory to replace it. Lakatos never gives a good or even a detailed argument for any of these propositions. Most of them are a consequence of a theory-bound philosophy and they are best revised or refuted by serious reflection on experimentation. I assess them in Part B, on Intervening. On crucial experiments and auxiliary hypotheses, see Chapter 15. On the distinctions between observation and theory, see Chapter 10.

Euclidean model and inductivism

In the beginning, says Lakatos, mathematical proof was the model of true science. Conclusions had to be demonstrated and made absolutely certain. Anything less than complete certainty was defective. Science was by definition infallible.

The seventeenth century and the experimental method of reasoning made this seem an impossible goal. Yet the tale is only modified as we pass from deduction to induction. If we cannot have secure knowledge let us at least have probable knowledge based on sure foundations. Observations rightly made shall serve as the basis. We shall generalize upon sound experiments, draw analogies, and build up to scientific conclusions. The greater the variety and quantity of observations that confirm a conclusion, the more probable it is. We may no longer have certainty, but we have high probability.

Here then are two stages on the high road to methodology: proof and probability. Hume, knowing the failure of the first, already cast doubts on the second by 1739. In no way can particular facts provide 'good reason' for more general statements or claims about the future. Popper agreed, and so in turn does Lakatos.

Falsificationisms

Lakatos truncates some history of methodology but expands others. He even had a Popper$_1$, Popper$_2$, and a Popper$_3$, denoting increasingly sophisticated versions of what Lakatos had learned from Popper. All three emphasize the testing and falsifying of conjectures rather than verifying or confirming them. The simplest view would be, 'people propose, nature disposes'. That is, we think up theories, and nature junks them if they are wrong. That implies a pretty sharp distinction between fallible theories and basic observations of nature. The latter, once checked out, are a final and indubitable court of appeal. A theory inconsistent with an observation must be rejected.

This story of conjecture and refutation makes us think of a pleasingly objective and honest science. But it won't do: for one thing 'all theories are born refuted', or at least it is very common for a theory to be proposed even when it is known not to square with all the known facts. That was Kuhn's point about puzzle-solving normal science. Secondly (according to Lakatos), there is no firm theory–observation distinction. Thirdly there is a claim made by the great French historian of science, Pierre Duhem. He remarked that theories are tested via auxiliary hypotheses. In his example, if an astronomer predicts that a heavenly body is to be found in a certain location, but it turns up somewhere else, he need not revise his astronomy. He could perhaps revise the theory of the telescope (or produce a suitable account of how phenomena differ from reality (Kepler), or invent a theory of astronomical aberration (G.G. Stokes), or suggest that the Doppler effect works differently in outer space). Hence a recalcitrant observation does not necessarily refute a theory. Duhem probably thought that it is a matter of choice or convention whether a theory or one of its auxiliary hypotheses is to be revised. Duhem was an outstanding anti-realist, so such a conclusion was attractive. It is repugnant to the staunch instincts for scientific realism found in Popper or Lakatos.

So the falsificationist adds two further props. First, no theory is rejected or abandoned unless there is a better rival theory in existence. Secondly, one theory is better than another if it makes more novel predictions. Traditionally theories had to be consistent with the evidence. The falsificationist, says Lakatos, demands not that the theory should be consistent with the evidence, but that it should actually outpace it.

Note that this last item has a long history of controversy. By and large inductivists think that evidence consistent with a theory supports it, no matter whether the theory preceded the evidence or the evidence preceded the theory. More rationalistic and deductively oriented thinkers will insist on what Lakatos calls 'the Leibniz–Whewell–Popper requirement that *the – well planned – building of pigeon holes must proceed much faster than the recording of facts which are to be housed in them*' (I, p. 100).

Research programmes

We might take advantage of the two spellings of the word, and use the American spelling 'research program' to denote what investigators normally call a research program, namely a specific attack on a problem using some well-defined combination of theoretical and experimental ideas. A research program is a program of research which a person or group can undertake, seek funding for, obtain help with, and so on. What Lakatos spells as 'research programme' is not much like that. It is more abstract, more historical. It is a sequence of developing theories that might last for centuries, and which might sink into oblivion for 80 years and then be revived by an entirely fresh infusion of facts or ideas.

In particular cases it is often easy to recognize a continuum of developing theories. It is less easy to produce a general characterization. Lakatos introduces the word 'heuristic' to help. Now 'heuristic' is an adjective describing a method or process that guides discovery or investigation. From the very beginnings of Artificial Intelligence in the 1950s, people spoke of heuristic procedures that would help machines solve problems. In *How to solve it* and other wonderful books, Lakatos's countryman and mentor, the mathematician Georg Polya, provided classic modern works on mathematical heuristics. Lakatos's work on the philosophy of mathematics owed much to Polya. He then adapted the idea of heuristics as a key to identifying research programmes. He says a research programme is defined by its positive and negative heuristic. The negative heuristic says: Hands off – don't meddle here. The positive heuristic says: Here is a set of problem areas ranked in order of importance – worry only about questions at the top of the list.

Hard cores and protective belts

The negative heuristic is the 'hard core' of a programme, a body of central principles which are never to be challenged. They are regarded as irrefutable. Thus in the Newtonian programme, we have at the core the three laws of dynamics and the law of gravitation. If planets misbehave, a Newtonian will not revise the gravitational law, but try to explain the anomaly by postulating a possibly invisible planet, a planet which, if need be, can be detected only by its perturbations on the solar system.

The positive heuristic is an agenda determining which problems are to be worked on. Lakatos imagines a healthy research programme positively wallowing in a sea of anomalies, but being none the less exuberant. According to him Kuhn's vision of normal science makes it almost a chance affair which anomalies are made the object of puzzle-solving activity. Lakatos says on the contrary that there is a ranking of problems. A few are systematically chosen for research. This choice generates a 'protective belt' around the theory, for one attends only to a set of problems ordained in advance. Other seeming refutations are simply ignored. Lakatos uses this to explain, why, *pace* Popper, verification seems so important in science. People choose a few problems to work on, and feel vindicated by a solution; refutations, on the other hand, may be of no interest.

Progress and degeneration

What makes a research programme good or bad? The good ones are progressive, the bad ones are degenerating. A programme will be a sequence of theories $T_1, T_2, T_3. . . .$ Each theory must be at least as consistent with known facts as its predecessor. The sequence is theoretically progressive if each theory in turn predicts some novel facts not foreseen by its predecessors. It is empirically progressive if some of these predictions pan out. A programme is simply *progressive*, if it is both theoretically and empirically progressive. Otherwise it is *degenerating*.

The degenerating programme is one that gradually becomes closed in on itself. Here is an example.[2] One of the famous success stories is that of Pasteur, whose work on microbes enabled him to save the French beer, wine and silk industries that were threatened by various small hostile organisms. Later we began to pasteurize milk. Pasteur also identified the micro-organisms that enabled him to vaccinate against anthrax and rabies. There evolved a research programme whose hard core held that every hitherto organic harm not explicable in terms of parasites or injured organs was to be explained in terms of micro-organisms. When many diseases failed to be caused by bacteria, the positive heuristic directed a search for something smaller, the virus. This progressive research programme had degenerating subprogrammes. Such was the enthusiasm for microbes that what we now call deficiency diseases *had* to be caused by bugs. In the early years of this century the leading professor of tropical disease, Patrick Manson, insisted that beriberi and some other deficiency diseases are caused by bacterial contagion. An

2 K. Codell Carter, 'The germ theory, Beri-beri, and the deficiency theory of disease', *Medical History* 21 (1977), pp. 119–36.

epidemic of beriberi was in fact caused by the new processes of steam-polishing rice, processes imported from Europe which killed off millions of Chinese and Indonesians whose staple food was rice. Vitamin B$_1$ in the hull of the rice was destroyed by polishing. Thanks largely to dietary experiments in the Japanese Navy, people gradually came to realize that not presence of microbes, but absence of something in polished rice was the problem. When all else failed, Manson insisted that there are bacteria that live and die in the polished but not in the unpolished rice, and they are the cause of the new scourge. This move was theoretically degenerating because each modification in Manson's theory came only after some novel observations, not before, and it was empirically degenerating because no polished-rice-organisms are to be found.

Hindsight

We cannot tell whether a research programme is progressive until after the fact. Consider the splendid problem shift of the Pasteur programme, in which viruses replace bacteria as the roots of most evils that persist in the developed world. In the 1960s arose the speculation that cancers – carcinomas and lymphomas – are caused by viruses. A few extremely rare successes have been recorded. For example, a strange and horrible tropical lymphoma (Burkitt's lymphoma) that causes grotesque swellings in the limbs of people who live above 5000 feet near the equator, has almost certainly been traced to a virus. But what of the general cancer-virus programme? Lakatos tells us, 'We must take budding programmes leniently; programmes may take decades before they get off the ground and become empirically progressive' (I, p. 6). Very well, but even if they have been progressive in the past – what more so than Pasteur's programme – that tells us exactly nothing except 'Be open-minded, and embark on numerous different kinds of research if you are stymied.' It does not merely fail to help choose new programmes with no track record. We know of few more progressive programmes than that of Pasteur, even if some of its failures have been hived off, for example into the theory of deficiency diseases. Is the attempt to find cancer viruses progressive or degenerating? We shall know only later. If we were trying to decide what proportion of the 'War on Cancer' to spend on molecular biology and what on viruses (not necessarily mutually exclusive, of course) Lakatos could tell us nothing.

Objectivity and subjectivism

What then was Lakatos doing? My guess is indicated by the title of this chapter. He wanted to find a substitute for the idea of truth. This is a little like Putnam's subsequent suggestion, that the correspondence theory of truth is mistaken, and truth is whatever it is rational to believe. But Lakatos is more radical than Putnam. Lakatos is no born-again pragmatist. He is down on truth, not just a particular theory of truth. He does not want a replacement for the correspondence theory, but a replacement for truth itself. Putnam has to fight himself away from a correspondence theory of truth because, in English-speaking philosophy, correspondence theories, despite the pragmatist assault of long ago, are still popular. Lakatos, growing up in an Hegelian tradition, almost never gives the correspondence theory a thought. However, like Peirce, he values an objectivity in science that plays little role in Hegelian discourse. Putnam honours this value by hoping, like Peirce, that there is a scientific method upon which we shall come to agree, and which in turn will lead us all to agreement, to rational, warranted, belief. Putnam is a simple Peircian, even if he is less confident than Peirce that we are already on the final track. Rationality looks forward. Lakatos went one step further. There is no forward-looking rationality, but we can comprehend the objectivity of our present beliefs by reconstructing the way we got here. Where do we start? With the growth of knowledge itself.

The growth of knowledge

The one fixed point in Lakatos's endeavour is the simple fact that knowledge does grow. Upon this he tries to build his philosophy without representation, starting from the fact that one can see that knowledge grows whatever we think about 'truth' or 'reality'. Three related aspects of this fact are to be noticed.

First, one can see by direct inspection that knowledge has grown. This is not a lesson to be taught by general philosophy or history but by detailed reading of specific sequences of texts. There is no doubt that more is known now than was grasped by past genius. To take an example of his own, it is manifest that after the work of Rutherford and Soddy and the discovery of isotopes, vastly more was known about atomic weights than had been dreamt of by a century of toilers after Prout had hypothesized in 1815 that hydrogen is the stuff of the universe, and that atomic weights are integral multiples of that of hydrogen. I state this to remind ourselves that Lakatos starts from a profound but elementary point. The point is not that there is knowledge but that there is growth; we know more about atomic weights than we once did, even if future times plunge us into quite new, expanded, reconceptualizations of those domains.

Secondly, there is no *arguing* that some historical events do exhibit the growth of knowledge. What is needed is an *analysis* that will say in what this growth consists, and tell us what is the growth that we call science and what is not. Perhaps there are fools who think that the discovery of isotopes is no growth in real knowledge. Lakatos's attitude is that they are not to be contested – they are likely idle and have never read the texts or engaged in the experimental results of such growth. We should not argue with such ignoramuses. When they have learned how to use isotopes or simply read the texts, they will find out that knowledge does grow.

This thought leads to the third point. The growth of scientific knowledge, given an intelligent analysis, might provide a demarcation between rational activity and irrationalism. Although Lakatos expressed matters in that way, it is not the right form of words to use. Nothing has grown more consistently and persistently over the years than the commentaries on the Talmud. Is that a rational activity? We see at once how hollow is that word 'rational' if used for positive evaluation. The commentaries are the most reasoned great bodies of texts that we know, vastly more reasoned than the scientific literature. Philosophers often pose the tedious question of why twentieth-century Western astrology, such as it is, is no science. That is not where the thorny issues of demarcation lie. Popper took on more serious game in challenging the right of psychoanalysis or Marxist historiography to the claim of 'science'. The machinery of research programmes, hard cores and protective belts, progress and degeneration, must, if it is of worth, effect a distinction not between the rational and reasoning, and the irrational and unreasoning, but between those reasonings which lead to what Popper and Lakatos call objective knowledge and those which pursue different aims and have different intellectual trajectories.

Appraising scientific theories

Hence Lakatos provides no forward-looking assessments of present competing scientific theories. He can at best look back and say why, on his criteria, this research programme was progressive, why another was not. As for the future, there are few pointers to be derived from his 'methodology'. He says that we should be modest in our hopes for our own projects because rival programmes may turn out to have the last word. There is a place for pig-headedness when one's programme is going through a bad patch. The mottos are to be proliferation of theories, leniency in evaluation, and honest 'score-keeping' to see which programme is producing results and meeting new challenges. These are not so much real methodology as a list of the supposed values of a science allegedly free of ideology.

If Lakatos were in the business of theory appraisal, then I should have to agree with his most colourful critic, Paul Feyerabend. The main thrust of the often perceptive assaults on Lakatos to be found in Chapter 17 of *Against Method* is that Lakatos's 'methodology' is not a good device for advising on current scientific work. I agree, but suppose that was never the point of the analysis which, I claim, has a more radical object. Lakatos had a sharp tongue, strong opinions and little diffidence. He made many entertaining observations about this or that current research project, but these acerbic asides were incidental to and independent of the philosophy I attribute to him.

Is it a defect in Lakatos's methodology that it is only retroactive? I think not. There are no significant general laws about what, in a current bit of research, bodes well for the future. There are only truisms. A group of workers who have just had a good idea often spends at least a few more years fruitfully applying it. Such groups properly get lots of money from corporations, governments, and foundations. There are other mild sociological inductions, for example that when a group is increasingly concerned to defend itself against criticism, and won't dare go out on a new limb, then it seldom produces interesting new research. Perhaps the chief practical problem is quite ignored by philosophers of rationality. How do you stop funding a program you have supported for five or fifteen years – a program to which many young people have dedicated their careers – and which is finding out very little? That real-life crisis has little to do with philosophy.

There is a current vogue among some philosophers of science, that Lakatos might have called 'the new justificationism'. It produces whole books trying to show that a system of appraising theories can be built up out of rules of thumb. It is even suggested that governments should fund work in the philosophy of science, in order to learn how to fund projects in real science. We should not confuse such creatures of bureaucracy with Lakatos's attempt to understand the content of objective judgement.

Internal and external history

Lakatos's tool for understanding objectivity was something he called history. Historians of science, even those given to considerable flights of speculative imagination, find in Lakatos only 'an historical parody that makes one's hair stand on end'. That is Gerald Holton's characterization in *The Scientific Imagination* (p. 106); many colleagues agree.

Lakatos begins with an 'unorthodox, new demarcation between "internal" and "external" history' (I, p. 102), but is not very clear what is going on. External history commonly deals in economic, social and technological factors that are not directly involved in the content of a science, but which are deemed to influence or explain some events in the history of knowledge. External history might include an event like the first Soviet satellite to orbit the earth – Sputnik – which was followed by the instant investment of vast sums of American money in science education. Internal history is usually the history of ideas germane to the science, and attends to the motivations of research workers, their patterns of communication and lines of intellectual filiation – who learned what from whom.

Lakatos's internal history is to be one extreme on this spectrum. It is to exclude anything in the subjective or personal domain. What people believed is irrelevant: it is to be a history of some sort of abstraction. It is, in short, to be a history of Hegelian alienated knowledge, the history of anonymous and autonomous research programmes.

This idea about the growth of knowledge into something

objective and non-human was foreshadowed in his first major philosophical work, *Proofs and Refutations*. On p. 146 of this wonderful dialogue on the nature of mathematics, we find:

Mathematical activity is human activity. Certain aspects of this activity – as of any human activity – can be studied by psychology, others by history. Heuristic is not primarily interested in these aspects. But mathematical activity produces mathematics. Mathematics, this product of human activity, 'alienates itself' from the human activity which has been producing it. It becomes a living growing organism that acquires a certain autonomy from the activity which has produced it.

Here then are the seeds of Lakatos's redefinition of 'internal history', the doctrine underlying his 'rational reconstructions'. One of the lessons of *Proofs and Refutations* is that mathematics might be both the product of human activity and autonomous, with its own internal characterization of objectivity which can be analysed in terms of how mathematical knowledge has grown. Popper has suggested that such objective knowledge could be a 'third world' of reality, and Lakatos toyed with this idea.

Popper's metaphor of a third world is puzzling. In Lakatos's definition, 'the "first world" is the physical world; the "second world" is the world of consciousness, of mental states and, in particular, of beliefs; the "third world" is the Platonic world of objective spirit, the world of ideas' (II, p. 108). I myself prefer those texts of Popper's where he says that the third world is a world of books and journals stored in libraries, of diagrams, tables and computer memories. Those extra-human things, uttered sentences, are more real than any talk of Plato would suggest.

Stated as a list of three worlds we have a mystery. Stated as a sequence of three emerging kinds of entity with corresponding laws it is less baffling. First there was the physical world. Then when sentient and reflective beings emerged out of that physical world there was also a second world whose descriptions could not be in any general way reduced to physical world descriptions. Popper's third world is more conjectural. His idea is that there is a domain of human knowledge (sentences, print-outs, tapes) which is subject to its own descriptions and laws and which cannot be reduced to second-world events (type by type) any more than second-world events can be reduced to first-world ones. Lakatos persists in the metaphorical expression of this idea: 'The *products* of human knowledge; propositions, theories, systems of theories, problems, problemshifts, research programmes live and grow in the "third world"; the producers of knowledge live in the first and second worlds' (II, p. 108). One need not be so metaphorical. It is a difficult but straightforward question whether there is an extensive and coherent body of description of 'alienated' and autonomous human knowledge that cannot be reduced to histories and psychologies of subjective beliefs. A substantiated version of a 'third world' theory can provide just the domain for the content of mathematics. It admits that mathematics is a product of the human mind, and yet is also autonomous of anything peculiar to psychology. An extension of this theme is provided by Lakatos's conception of 'unpsychological' internal history.

Internal history will be a rational construction of what actually happened, one which displays why what happened in many of the best incidents of the history of science are worthy of designations such as 'rational' and 'objective'. Lakatos had a fine sounding maxim, a parody of one of Kant's noble turns of phrase: 'Philosophy of science without history of science is empty; history of science without philosophy of science is blind.' That sounds good, but Kant had been speaking of something else. All we need to say about rather unreflective history of science was said straightforwardly by Kant himself in his lectures on *Logic*: 'Mere polyhistory is a *cyclopean* erudition that lacks one eye, the eye of philosophy.' Lakatos wants to rewrite the history of science so that the 'best' incidents in the history of science are cases of progressive research programmes.

Rational reconstruction

Lakatos has a problem, to characterize the growth of knowledge internally by analysing examples of growth. There is a conjecture, that the unit of growth is the research programme (defined by hard core, protective belt, heuristic) and that research programmes are progressive or degenerating and, finally, that knowledge grows by the triumph of progressive programmes over degenerating ones. To test this supposition we select an example which must prima facie illustrate something that scientists have found out. Hence the example should be currently admired by scientists, or people who think about the appropriate branch of knowledge, not because we kow-tow to orthodoxy, but because workers in a given domain tend to have a better sense of what matters than laymen. Feyerabend calls this attitude elitism. Is it? The next Lakatosian injunction is for all of us to read all the texts we can lay hands on, covering a complete epoch spanned by the research programme, and the entire array of practitioners. Yes, that is elitism because few can afford the time to read. But it has an anti-elite intellectual premise (as opposed to an elite economic premise) that if texts are available, anyone is able to read them.

Within what we read we must select the class of sentences that express what the workers of the day were trying to find out, and how they were trying to find it out. Discard what people felt about it, the moments of creative hype, even their motivation or their role models. Having settled on such an 'internal' part of the data we can now attempt to organize the result into a story of Lakatosian research programmes.

As in most inquiries, an immediate fit between conjecture and articulated data is not to be expected. Three kinds of revision may improve the mesh between conjecture and selected data. First, we may fiddle with the data analysis, secondly, we may revise the conjecture, and thirdly, we may conclude that our chosen case study does not, after all, exemplify the growth of knowledge. I shall discuss these three kinds of revision in order.

By improving the analysis of data I do not mean lying. Lakatos made a couple of silly remarks in his 'falsification' paper, where he asserts something as historical fact in the text, but retracts it in the footnotes, urging that we take his text with tons of salt (I, p. 55). The historical reader is properly irritated by having his nose tweaked in this way. No point was being served. Lakatos's little joke was not made in the course of a rational reconstruction despite the fact that he said it was. Just as in any other inquiry, there is nothing wrong with trying to re-analyse the data. That does not mean lying. It may mean simply reconsidering or selecting and arranging the facts, or it may be a case of imposing a new research programme on the known historical facts.

If the data and the Lakatosian conjecture cannot be reconciled, two options remain. First, the case history may itself be regarded as something other than the growth of knowledge. Such a gambit could easily become monster-barring, but that is where the constraint of external history enters. Lakatos can always say that a particular incident in the history of science fails to fit his model because it is 'irrational', but he imposes on himself the demand that one should allow this only if one can say what the irrational element is. External elements may be political pressure, corrupted values or, perhaps, sheer stupidity. Lakatos's histories are normative in that he can conclude that a given chunk of research 'ought not to have' gone the way it did, and that it went that way through the interference of external factors not germane to the programme. In concluding that a chosen case was not 'rational' it is permissible to go against current scientific wisdom. But although in principle Lakatos can countenance this, he is properly moved by respect for the implicit appraisals of working scientists. I cannot see Lakatos willingly conceding that Einstein, Bohr, Lavoisier or even Copernicus was participating in an irrational programme. 'Too much of the actual history of science' would then become 'irrational' (I, p. 172). We have no standards to appeal to, in Lakatos's programme, other than the history of knowledge as it stands. To declare it to be globally irrational is to abandon rationality. We see why Feyerabend spoke of Lakatos's elitism. Rationality will simply be defined by what a present community calls good, and nothing shall counterbalance the extraterrestrial weight of an Einstein.

Lakatos then defines objectivity and rationality in terms of progressive research programmes, and allows an incident in the history of science to be objective and rational if its internal history can be written as a sequence of progressive problem shifts.

Cataclysms in reasoning

Peirce defined truth as what is reached by an ideal end to scientific inquiry. He thought that it is the task of methodology to characterize the principles of inquiry. There is an obvious problem: what if inquiry should not converge on anything? Peirce, who was as familiar in his day with talk of scientific revolutions as we are in ours, was determined that 'cataclysms' in knowledge (as he called them) have not been replaced by others, but this is all part of the self-correcting character of inquiry. Lakatos has an attitude similar to Peirce's. He was determined to refute the doctrine that he attributed to Kuhn, that knowledge changes by irrational 'conversions' from one paradigm to another.

As I said in the Introduction, I do not think that a correct reading of Kuhn gives quite the apocalyptic air of cultural relativism that Lakatos found there. But there is a really deep worry underlying Lakatos's antipathy to Kuhn's work, and it must not be glossed over. It is connected with an important side remark of Feyerabend's, that Lakatos's accounts of scientific rationality at best fit the major achievements 'of the last couple of hundred years'.

A body of knowledge may break with the past in two distinguishable ways. By now we are all familiar with the possibility that new theories may completely replace the conceptual organization of their predecessors. Lakatos's story of progressive and degenerating programmes is a good stab at deciding when such replacements are 'rational'. But all of Lakatos's reasoning takes for granted what we may call the hypothetico-deductive model of reasoning. For all his revisions of Popper, he takes for granted that conjectures are made and tested against some problems chosen by the protective belt. A much more radical break in knowledge occurs when an entirely new style of reasoning surfaces. The force of Feyerabend's gibe about 'the last couple of hundred years' is that Lakatos's analysis is relevant not to timeless knowledge and timeless reason, but to a particular kind of knowledge produced by a particular style of reasoning. That knowledge and that style have specific beginnings. So the Peircian fear of cataclysm becomes: Might there not be further styles of reasoning which will produce yet a new kind of knowledge? Is not Lakatos's surrogate for truth a local and recent phenomenon?

I am stating a worry, not an argument. Feyerabend makes sensational but implausible claims about different modes of reasoning and even seeing in the archaic past. In a more pedestrian way my own book, *The Emergence of Probability* (1975), contends that part of our present conception of inductive evidence came into being only at the end of the Renaissance. In his book, *Styles of Scientific Thinking in the European Tradition* (1983), the historian A.C. Crombie, from whom I take the word 'style', writes of six distinguishable styles. I have elaborated Crombie's idea elsewhere. Now it does not follow that the emergence of a new style is a cataclysm. Indeed we may add style to style, with a cumulative body of conceptual tools. That is what Crombie teaches. Clearly both

Putnam and Laudan expect this to happen. But these are matters which are only recently broached, and are utterly ill-understood. They should make us chary of an account of reality and objectivity which starts from the growth of knowledge, when the kind of growth described turns out to concern chiefly a particular knowledge achieved by a particular style of reasoning.

To make matters worse, I suspect that a style of reasoning may determine the very nature of the knowledge that it produces. The postulational method of the Greeks gave a geometry which long served as the philosopher's model of knowledge. Lakatos inveighs against the domination of the Euclidean mode. What future Lakatos will inveigh against the hypothetico-deductive mode and the theory of research programmes to which it has given birth? One of the most specific features of this mode is the postulation of theoretical entities which occur in high-level laws, and yet which have experimental consequences. This feature of successful science becomes endemic only at the end of the eighteenth century. Is it even possible that the questions of objectivity, asked for our times by Kant, are precisely the questions posed by this new knowledge? If so, then it is entirely fitting that Lakatos should try to answer those questions in terms of the knowledge of the past two centuries. But it would be wrong to suppose that we can get from this specific kind of growth to a theory of truth and reality. To take seriously the title of a book that Lakatos proposed, but never lived to write, 'The changing logic of scientific discovery' is to take seriously the possibility that Lakatos has, like the Greeks, made the eternal verities depend on a mere episode in the history of human knowledge.

There remains an optimistic version of this worry. Lakatos was trying to characterize certain objective values of Western science without an appeal to copy theories of truth. Maybe those objective values are recent enough that his limitation to the past two or three centuries is exactly right. We are left with no external way to evaluate our own tradition, but why should we want that?

BREAK

BREAK

BREAK
Reals and representations

Incommensurability, transcendental nominalism, surrogates for truth, and styles of reasoning are the jargon of philosophers. They arise from contemplating the connection between theory and the world. All lead to an idealist cul-de-sac. None invites a healthy sense of reality. Indeed much recent philosophy of science parallels seventeenth-century epistemology. By attending only to knowledge as representation of nature, we wonder how we can ever escape from representations and hook-up with the world. That way lies an idealism of which Berkeley is the spokesman. In our century John Dewey has spoken sardonically of a spectator theory of knowledge that has obsessed Western philosophy. If we are mere spectators at the theatre of life, how shall we ever know, on grounds internal to the passing show, what is mere representation by the actors, and what is the real thing? If there were a sharp distinction between theory and observation, then perhaps we could count on what is observed as real, while theories, which merely represent, are ideal. But when philosophers begin to teach that all observation is loaded with theory, we seem completely locked into representation, and hence into some version of idealism.

Pity poor Hilary Putnam, for example. Once the most realist of philosophers, he tried to get out of representation by tacking 'reference' on at the end of the list of elements that constitute the meaning of a word. It was as if some mighty referential sky-hook could enable our language to embed within it a bit of the very stuff to which it refers. Yet Putnam could not rest there, and ended up as an 'internal realist' only, beset by transcendental doubts, and given to some kind of idealism or nominalism.

I agree with Dewey. I follow him in rejecting the false dichotomy between acting and thinking from which such idealism arises. Perhaps all the philosophies of science that I have described are part of a larger spectator theory of knowledge. Yet I do not think that the idea of knowledge as representation of the world is in itself the source of that evil. The harm comes from a single-minded obsession with representation and thinking and theory, at the expense of intervention and action and experiment. That is why in the next part of this book I study experimental science, and find in it the sure basis of an uncontentious realism. But before abandoning theory for experiment, let us think a little more about the very notions of representation and reality.

The origin of ideas

What are the origins of these two ideas, *representation* and *reality*? Locke might have asked that question as part of a psychological inquiry, seeking to show how the human mind forms, frames, or constitutes its ideas. There is a legitimate science that studies the maturation of human intellectual abilities, but philosophers often play a different game when they examine the origin of ideas. They tell fables in order to teach philosophical lessons. Locke himself was fashioning a parable when he pretended to practice the natural history of the mind. Our modern psychologies have learned how to trick themselves out in more of the paraphernalia of empirical research, but they are less distant from fantastical Locke than they assume. Let us, as philosophers, welcome fantasies. There may be more truth in the average *a priori* fantasy about the human mind than in the supposedly disinterested observations and mathematical model-building of cognitive science.

Philosophical anthropology

Imagine a philosophical text of about 1850: 'Reality is as much an anthropomorphic creation as God Himself.' This is not to be uttered in a solemn tone of voice that says, 'God is dead and so is reality.' It is to be a more specific and practical claim: *Reality is just a byproduct of an anthropological fact.* More modestly, the concept of reality is a byproduct of a fact about human beings.

By anthropology I do not mean ethnography or ethnology, the studies practised in present-day departments of anthropology, and which involve lots of field work. By anthropology I mean the bogus nineteenth-century science of 'Man'. Kant once had three philosophical questions. What must be the case? What should we do? For what may we hope? Late in life he added a fourth question: *What is Man?* With this he inaugurated (*philosophische*) *Anthropologie* and even wrote a book called *Anthropology*. Realism is not to be considered part of pure reason, nor judgement, nor the metaphysics of morals, nor even the metaphysics of natural science. If we are to give it classification according to the titles of Kant's great books, realism shall be studied as part of *Anthropologie* itself.

A Pure Science of Human Beings is a bit risky. When Aristotle proposed that Man is an animal that lives in cities, so that the *polis* is a part of Man's nature to which He strives, his pupil Alexander refuted him by re-inventing the Empire. We have been told that Man is a tool-maker, or a creature that has a thumb, or that stands erect. We have been told that these fortuitous features are noticed only by attending to half of the species wrongly called Man, and that tools, thumbs and erectness are scarcely what define the race. It is seldom clear what the grounds might be for any such statements, pro or con. Suppose one person defines humans as rational, and another person defines them as the makers of tools. Why on earth should we suppose that being a rational animal is co-extensive with making tools?

Speculations about the essential nature of humanity license more of the same. Philosophers since Descartes have been attracted by the conjecture that humans are speakers. It has been urged that rationality, of its very nature, demands language, so humans as rational animals, and humans as speakers are indeed co-extensive. That is a satisfactory main theorem for a subject as feeble as fanciful anthropology. Yet despite the manifest profundity of this conclusion, a conclusion that has fuelled mighty books, I propose another fancy. *Human beings are representers.* Not *homo faber*, I say, but *homo depictor*. People make representations.

Limiting the metaphor

People make likenesses. They paint pictures, imitate the clucking of hens, mould clay, carve statues, and hammer brass. Those are the sorts of representations that begin to characterize human beings.

The word 'representation' has quite a philosophical past. It has been used to translate Kant's word *Vorstellung*, a placing before the mind, a word which includes images as well as more abstract thoughts. Kant needed a word to replace the 'idea' of the French and English empiricists. That is exactly what I do *not* mean by representation. Everything I call a representation is public. You cannot touch a Lockeian idea, but only the museum guard can stop you touching some of the first representations made by our predecessors. I do not mean that all representations can be touched, but all are public. According to Kant, a judgement is a representation of a representation, a putting before the mind of a putting before the mind, doubly private. That is doubly not what I call a representation. But for me, some public verbal events can be representations. I think not of simple declarative sentences, which are surely not representations, but of complicated speculations which attempt to represent our world.

When I speak of representations I first of all mean physical objects: figurines, statues, pictures, engravings, objects that are themselves to be examined, regarded. We find these as far back as we find anything human. Occasionally some fortuitous event preserves even fragments of wood or straw that would otherwise have rotted. Representations are external and public, be they the simplest sketch on a wall, or, when I stretch the word 'representation', the most sophisticated theory about electromagnetic, strong, weak, or gravitational forces.

The ancient representations that are preserved are usually visual and tactile, but I do not mean to exclude anything publicly accessible to the other senses. Bird whistles and wind machines may make likenesses too, even though we usually call the sounds that they emit imitations. I claim that if a species as smart as human beings had been irrevocably blind, it would have got on fine with auditory and tactile representations, for to represent is part of our very nature. Since we have eyes, most of the first representations were visual, but representation is not of its essence visual.

Representations are intended to be more or less public likenesses. I exclude Kant's *Vorstellungen* and Lockeian internal ideas that represent the external world in the mind's eye. I also exclude ordinary public sentences. William James jeered at what he called the copy theory of truth, which bears the more dignified label of correspondence theory of truth. The copy theory says that true propositions are copies of whatever in the world makes them true. Wittgenstein's *Tractatus* has a picture theory of truth, according to which a true sentence is one which correctly pictures the facts. Wittgenstein was wrong. Simple sentences are not pictures, copies, or representations. Doubtless philosophical talk of representation invites memories of Wittgenstein's *Sätze*. Forget them. The sentence, 'the cat is on the mat', is no representation of reality. As Wittgenstein later taught us, it is a sentence that can be used for all sorts of purposes, none of which is to portray what the world is like. On the other hand, Maxwell's electromagnetic theories were intended to represent the world, to say what it is like. Theories, not individual sentences, are representations.

Some philosophers, realizing that sentences are not representations, conclude that the very idea of a representation is worthless for philosophy. That is a mistake. We can use complicated sentences collectively in order to represent. So much is ordinary English idiom. A lawyer can represent the client, and can also represent that the police collaborated improperly in preparing their reports. A *single* sentence will in general not represent. A representation can be verbal, but a verbal representation will use a good many verbs.

Humans as speakers

The first proposition of my philosophical anthropology is that human beings are depictors. Should the ethnographer tell me of a race that makes no image (not because that is tabu but because no one has thought of representing anything) then I would have to say that those are not people, not *homo depictor*. If we are persuaded that humankind (and not its predecessors) lived in Olduvai gorge three million years ago, and yet we find nothing much except old skulls and footprints, I would rather postulate that the representations made by those African forbears have been erased by sand, rather than that people had not yet begun to represent.

How does my *a priori* paleolithic fantasy mesh with the ancient idea that humans are essentially rational and that rationality is essentially linguistic? Must I claim that depiction needs language or that humanity need not be rational? If language has to be tucked into rationality, I would cheerfully conclude that humans may *become* rational animals. That is, *homo depictor* did not always deserve Aristotle's accolade of rationality, but only earned it as we smartened up and began to talk. Let us imagine, for a moment, pictorial people making likenesses before they learn to talk.

The beginnings of language

Speculation on the origin of language tends to be unimaginative and condescending. Language, we hear, must have been invented to help with practical matters such as hunting and farming. 'How useful,' goes the refrain, 'to be able to talk. How much more efficient people would have been if they could talk. Speech makes it much more likely that hunters and farmers will survive.'

Scholars who favour such rubbish have evidently never ploughed a field nor stalked game, where silence is the order of the day, not jabber. People out in the fields weeding do not usually talk. They talk only when they rest. In the plains of East Africa the hunter with the best kill rate is the wild dog, yet middle-aged professors short of wind and agreeing never to talk nor signal are much better at catching the beeste and the gazelle than any wild dog. The lion that roars and the dogs that bark will starve to death if enough silent humans are hunting with their bare hands.

Language is not for practical affairs. Jonathan Bennett tells a story about language beginning when one 'tribesman' warns another that a coconut is about to fall on the second native's head.[1] Native One does this first by an overacted mime of bonking on the head, and later on does this by uttering a warning and thereby starting language. I bet that no coconut ever fell on any tribesman's head except in racist comic strips, so I doubt this fantasy. I prefer a suggestion about language attributed to the Leakey family who excavate Olduvai gorge. The idea is that people invented language out of boredom. Once we had fire, we had nothing to do to pass away the long evenings, so we started telling jokes. This fancy about the origin of language has the great merit of regarding speech as something human. It fixates not on tribesmen in the tropics but on people.

Imagine *homo depictor* beginning to use sounds that we might translate as 'real', or, 'that's how it is', said of a clay figurine or a daub on the wall. Let discourse continue as 'this real, then that real', or, more idiomatically, 'if this is how it is, then that is how it is too'. Since people are argumentative, other sounds soon express, 'no, not that, but this here is real instead'.

1 J. Bennett, 'The meaning-nominalist strategy', *Foundations of Language* 10 (1973), pp. 141–68.

In such a fantasy we do not first come to the names and descriptions, or the sense and reference of which philosophers are so fond. Instead we start with the indexicals, logical constants, and games of seeking and finding. Descriptive language comes later, not as a surrogate for depiction but as other uses for speaking are invented.

Language then starts with 'this real', said of a representation. Such a story has to its credit the fact that 'this real' is not at all like 'You Tarzan, Me Jane', for it stands for a complicated, that is, characteristically human, thought, namely that this wooden carving shows something real about what it represents.

This imagined life is intended as an antidote to the deflating character of the quotation with which I began: Reality is an anthropomorphic creation. Reality may be a human creation, but it is no toy; on the contrary it is the second of human creations. The first peculiarly human invention is representation. Once there is a practice of representing, a second-order concept follows in train. This is the concept of reality, a concept which has content only when there are first-order representations.

It will be protested that reality, or the world, was there before any representation or human language. Of course. But conceptualizing it as reality is secondary. First there is this human thing, the making of representations. Then there was the judging of representations as real or unreal, true or false, faithful or unfaithful. Finally comes the world, not first but second, third or fourth.

In saying that reality is parasitic upon representation, I do not join forces with those who, like Nelson Goodman or Richard Rorty, exclaim, 'the world well lost!' The world has an excellent place, even if not a first one. It was found by conceptualizing the real as an attribute of representations.

Is there the slightest empirical evidence for my tale about the origin of language? No. There are only straws in the wind. I say that representing is curiously human. Call it species specific. We need only run up the evolutionary tree to see that there is some truth in this. Drug a baboon and paint its face, then show it a mirror. It notices nothing out of the ordinary. Do the same to a chimpanzee. It is terribly upset, sees there is paint on its face and tries to get if off. People, in turn, like mirrors to study their make-up. Baboons will never draw pictures. The student of language, David Premack, has taught chimpanzees a sort of language using pictorial representation. *Homo depictor* was better than that, right from the start. We still are.

Likeness

Representations are first of all likenesses. Saying so flies in the face of philosophical truisms. There is, we all know, no representation without style. Even the most untutored of cultures must have a system of representation if it is to represent at all. So it may be argued that there cannot in the beginning have been simply representation, a creating of likeness. There must have been a style of representing before there was representing.

I need not disagree with this doctrine, so long as it be admitted that styles do not precede representation. They grow with representation as materials are worked, and craftspeople produce artifacts that affect the sensibility of their customers.

A more philosophical conundrum lurks hereabouts. Things are alike, it is said, in some respect or another, and cannot be simply like. There must be some concept used to express that in which likeness consists. Two people have the same walk, or the same bearing, or the same nose, or the same parents or the same character. But two people cannot simply be 'like' each other. I agree with this too, but tentatively hold that it does not preclude simple likeness.

I am too brainwashed by philosophy to hold that things *in general* can be simply, or unqualifiedly alike. They must be like or unlike in this or that respect. However a particular kind of thing, namely a human-made representation, can unqualifiedly be like what it is intended to represent. Our generalized notion of likeness is, like our idea of reality, parasitic upon our practices of representation. There may be some initial way in which representations are like what they represent. There is no doubt that some human artifacts of very foreign and very ancient peoples are immediately recognized as likenesses, even when we do not quite know what they are likenesses of. Those pictures, carvings, gold inlay, worked copper, clay faces, mammoth rock carvings, pocket-sized canoes for burial purposes – all the artistic detritus that we find where people once lived – are likenesses. I may not know what they are likenesses of nor what they are for. I ill-understand the systems of representation but I know these are representations all the same. At Delphi I see an archaic ivory carving of a person, perhaps a god, in what we call formal or lifeless style. I see the gold leggings and cloak in which the ivory was dressed. It is engraved in the most minute and 'realistic' detail with scenes of bull and lion. The archaic and the realistic objects in different media are made in what the archaeologists say is the same period. I do not know what either is for. I do know that both are likenesses. I see the archaic bronze charioteer with its compelling human deep-set eyes of semi-precious stone. How, I ask, could craftspeople so keen on what we call lifeless forms work with others who breathed life into their creations? Because different crafts using different media evolve at different rates? Because of a forgotten combination of unknown purposes? Such subtle questions are posed against a background of what we take for granted. We know at least this: these artifacts are representations.

We know likeness and representation even when we cannot answer, likeness to what? Think of the strange little clay figures on which are painted a sketch of garments, but which have, instead of heads, little saucer shaped depressions, perhaps for oil. These finger-high objects litter Mycenae. I doubt that they represent anything in particular. They most remind me of the angel-impressions children make by lying in the snow and waving their arms and legs to and fro to create the image of little wings and skirt. Children make these angels for pleasure. We do not quite know what the citizens of Cnossus did with their figurines. But we know that both are in some way likenesses. The wings and skirt are like wings and skirt, although the angel depicted is like nothing on earth.

Representations are not in general intended to say how it is. They can be portrayals or delights. After our recent obsession with words it is well to reflect on pictures and carvings. Philosophers of language seldom resist the urge to say that the first use of language must be to tell the truth. There should be no such compulsion with pictures. To argue of two bison sketches, 'If this is how it is, then that is how it is too', is to do something utterly unusual. Pictures are seldom, and statues are almost never used to say how things are. At the same time there is a core to representation that enables archaeologists millenia later to pick out certain objects in the debris of an ancient site, and to see them as likenesses. Doubtless 'likeness' is the wrong word, because the 'art' objects will surely include products of the imagination, pretties and uglies made for their own sake, for the sake of revenge, wealth, understanding, courtship or terror. But within them all there is a notion of representation that harks back to likeness. Likeness stands alone. It is not a relation. It creates the terms in a relation. There is first of all likeness, and then likeness to something or other. First there is representation, and then there is 'real'. First there is a representation and much later there is a creating of concepts in terms of which we can describe this or that respect in which we have similarity. But likeness can stand on its own without any need of some concepts x, y, or z, so that one must always think, like in represent of z, but not of x or y. There is no absurdity in thinking that there is a raw and unrefined notion of likeness springing up with the making of representations, and which, as people become more skilful in working with materials, engenders all sorts of different ways of noticing what is like what.

Realism no problem

If reality were just an attribute of representation, and we had not evolved alternative styles of representation, then realism would be a problem neither for philosophers nor for aesthetes. The problem arises because we have alternative systems of representation.

So much is the key to the present philosophical interest in scientific realism. Earlier 'realistic' crises commonly had their roots in science. The competition between Ptolemaic and Copernican systems begged for a shoot-out between instrumentalist and realistic cosmologies. Disputes about atomism at the end of the nineteenth century made people wonder if, or in what sense, atoms could be real. Our present debate about scientific realism is fuelled by no corresponding substantive issue in natural science. Where then does it come from? From the suggestions of Kuhn and others that with the growth of knowledge we may, from revolution to revolution, come to inhabit different worlds. New theories are new representations. They represent in different ways and so there are new kinds of reality. So much is simply a consequence of my account of reality as an attribute of representation.

When there were only undifferentiated representations then, in my fantasy story about the origin of language, 'real' was unequivocal. But as soon as representations begin to compete, we had to wonder what is real. Anti-realism makes no sense when only one kind of representation is around. Later it becomes possible. In our time we have seen this as the consequence of Kuhn's *Structure of Scientific Revolutions*. It is, however, quite an old theme in philosophy, best illustrated by the first atomists.

The Democritean dream

Once representation was with us, reality could not be far behind. It is an obvious notion for a clever species to cultivate. The prehistory of our culture is necessarily given by representations of various sorts, but all that are left us are tiny physical objects, painted pots, moulded cookware, inlay, ivory, wood, tiny burial tools, decorated walls, chipped boulders. *Anthropologie* gets past the phantasies I have constructed only when we have the remembered word, the epics, incantations, chronologies and speculations. The pre-Socratic fragments would be so much mumbo-jumbo were it not for their lineage down to the strategies we now calmly call 'science'. Today's scientific realist attends chiefly to what was once called the inner constitution of things, so I shall pull down only one thread from the pre-Socratic skein, the one that leads down to atomism. Despite Leucippus, and other forgotten predecessors, it is natural to associate this with Democritus, a man only a little older than Socrates. The best sciences of his day were astronomy and geometry. The atomists were bad at the first and weak in the second, but they had an extraordinary hunch. Things, they supposed, have an inner constitution, a constitution that can be thought about, perhaps even uncovered. At least they could guess at this: atoms and the void are all that exist, and what we see and touch and hear are only modifications of this.

Atomism is not essential to this dream of knowledge. What matters is an intelligible organization behind what we take in by the senses. Despite the central role of cosmology, Euclidean proof, medicine and metallurgy in the formation of Western culture, our current problems about scientific realism stem chiefly from the Democritean dream. It aims at a new kind of representation. Yet it still aims at likeness. This stone, I imagine a Democritus saying, is not as it looks to the eye. It is like this – and here he draws dots in the sand or on the tablet, itself thought of as a void. These dots are in continuous and uniform motion, he says, and begins to tell a tale of particles that his descendants turn into odd shapes, springs, forces, fields, all too small or big to be seen or felt or heard except in the aggregate. But the aggregate, continues Democritus, is none other than this stone, this arm, this earth, this universe.

Familiar philosophical reflections ensue. Scepticism is inevitable, for if the atoms and the void comprise the real, how can we ever know that? As Plato records in the *Gorgias*, this scepticism is three-pronged. All scepticism had had three prongs, since Democritus formulated atomism. There is first of all the doubt that we could check out any particular version of the Democritean dream. If much later Lucretius adds hooks to the atoms, how can we know if he or another speculator is correct? Secondly, there is a fear that this dream is only a dream; there are no atoms, no void, just stones, about which we can, for various purposes, construct certain models whose only touchstone, whose only basis of comparison, whose only reality, is the stone itself. Thirdly, there is the doubt that, although we cannot possibly believe Democritus, the very possibility of his story shows that we cannot credit what we see for sure, and so perhaps we had better not aim at knowledge but at the contemplative ignorance of the tub.

Philosophy is the product of knowledge, no matter how sketchy be the picture of what is known. Scepticism of the sort 'do I know this is a hand before me' is called 'naive' when it would be better described as degenerate. The serious scepticism which is associated with it is not, 'is this a hand rather than a goat or an hallucination?' but one that originates with the more challenging worry that the hand represented as flesh and bone is false, while the hand represented as atoms and the void is more correct. Scepticism is the product of atomism and other nascent knowledge. So is the philosophical split between appearance and reality. According to the Democritean dream, the atoms must be like the inner constitution of the stone. If 'real' is an attribute of depiction, then in asserting his doctrine, Democritus can only say that his picture of particles pictures reality. What then of the depiction of the stone as brown, encrusted, jagged, held in the hand? That, says the atomist, must be appearance.

Unlike its opposite, reality, 'appearance' is a thoroughly philosophical concept. It imposes itself on top of the initial two tiers of representation and reality. Much philosophy misorders this triad. Locke thought that we have appearance, then form mental representations and finally, seek reality. On the contrary, we make public representations, form the concept of reality, and, as systems of representation multiply, we become sceptics and form the idea of mere appearance.

No one calls Democritus a scientific realist: 'atomism' and 'materialism' are the only 'isms' that fit. I take atomism as the natural step from the Stone Age to scientific realism, because it lays out the notion of an 'inner constitution of things'. With this seventeenth-century phrase, we specify a constitution to be thought about and, hopefully, to be uncovered. But no one did find out about atoms for a long, long time. Democritus transmitted a dream, but no knowledge. Complicated concepts need criteria of application. That is what Democritus lacked. He did not know enough beyond his speculations to have criteria of whether his picture was of reality or not. His first move was to shout 'real' and slander the looks of things as mere appearance. Scientific realism or anti-realism do not become possible doctrines until there are criteria for judging whether the inner constitution of things is as represented.

The criteria of reality

Democritus gave us one representation: the world is made up of atoms. Less occult observers give us another. They painted pebbles on the beach, sculpted humans and told tales. In my account, the word 'real' first meant just unqualified likeness. But then clever people acquired conjectured likenesses in manifold respects. 'Real' no longer was unequivocal. As soon as what we would now call speculative physics had given us alternative pictures of reality, metaphysics was in place. Metaphysics is about criteria of reality. Metaphysics is intended to sort good systems of representation from bad ones. Metaphysics is put in place to sort representations when the only criteria for representations are supposed to be internal to representation itself.

That is the history of old metaphysics and the creation of the problem of realism. The new era of science seemed to save us from all that. Despite some philosophical malcontents like Berkeley, the new science of the seventeenth century could supplant even organized religion and say that it was giving the true representation of the world. Occasionally one got things wrong, but the overthrow of false ideas was only setting us on what was finally the right path. Thus the chemical revolution of Lavoisier was seen as a real revolution. Lavoisier got some things wrong: I have twice already used the example of his confidence that all acids have oxygen in them. So we sorted that out. In 1816 the new professor of chemistry at Harvard College relates the history of chemistry in an inaugural lecture to the teenagers then enrolled. He notes the revolutions of the recent past, and says we are now on the right road. From now on there will only be corrections. All of that was fine until it began to be realized that *there might be several ways to represent the same facts*.

I do not know when this idea emerged. It is evident in the important posthumous book of 1894, Heinrich Hertz's *Principles of Mechanics*. This is a remarkable work, often said to have led Wittgenstein towards his picture theory of meaning, the core of his 1918 *Tractatus Logico-Philosophicus*. Perhaps this book, or its 1899 English translation, first offers the explicit terminology of a scientific 'image' – now immortalized in the opening sentence of Kuhn's *Structure*, and, following Wilfred Sellars, used as the title of van Fraassen's anti-realist book. Hertz presents 'three images of mechanics' – three different ways to represent the then extant knowledge of the motions of bodies. Here, for perhaps the first time, we have three different systems of representation shown to us. Their merits are weighed, and Hertz favours one.

Hence even within the best understood natural science – mechanics – Hertz needed criteria for choosing between representations. It is not only the artists of the 1870s and 1880s who are giving us new systems of representation called post-impressionism or whatever. Science itself has to produce criteria of what is 'like', of what shall count as the right representation. Whereas art learns to live with alternative modes of representation, here is Hertz valiantly trying to find uniquely the right one for mechanics. None of the traditional values – values still hallowed in 1983 – values of prediction, explanation, simplicity, fertility, and so forth, quite do the job. The trouble is, as Hertz says, that all three ways of representing mechanics do a pretty good job, one better at this, one better at that. What then is the truth about the motions of bodies? Hertz invites the next generation of positivists, including Pierre Duhem, to say that there is no truth of the matter – there are only better or worse systems of representation, and there might well be inconsistent but equally good images of mechanics.

Hertz was published in 1894, and Duhem in 1906. Within that

span of years pretty well the whole of physics was turned upside down. Increasingly, people who knew no physics gossiped that everything is relative to your culture, but once again physicists were sure they were on the only path to truth. They had no doubt about the right representation of reality. We have only one measure of likeness: the hypothetico-deductive method. We propose hypotheses, deduce consequences and see if they are true. Hertz's warnings that there might be several representations of the same phenomena went unheeded. The logical positivists, the hypothetico-deductivists, Karl Popper's falsificationists – they were all deeply moved by the new science of 1905, and were scientific realists to a man, even when their philosophy ought to have made them somewhat anti-realist. Only at a time when physics was rather quiescent would Kuhn cast the whole story in doubt. Science is not hypothetico-deductive. It does have hypotheses, it does make deductions, it does test conjectures, but none of these determine the movement of theory. There are – in the extremes of reading Kuhn – no criteria for saying which representation of reality is the best. Representations get chosen by social pressures. What Hertz had held up as a possibility too scaring to discuss, Kuhn said was brute fact.

Anthropological summary

People represent. That is part of what it is to be a person. In the beginning to represent was to make an object like something around us. Likeness was not problematic. Then different kinds of representation became possible. What was like, which real? Science and its philosophy had this problem from the very beginning, what with Democritus and his atoms. When science became the orthodoxy of the modern world we were able, for a while, to have the fantasy that there is one truth at which we aim. That is the correct representation of the world. But the seeds of alternative representations were there. Hertz laid that out, even before the new wave of revolutionary science which introduced our own century. Kuhn took revolution as the basis for his own implied anti-realism. We should learn this: When there is a final truth of the matter – say, the truth that my typewriter is on the table – then what we say is either true or false. It is not a matter of representation. Wittgenstein's *Tractatus* is exactly wrong. Ordinary simple atomic sentences are not representations of anything. If Wittgenstein derived his picture account of meaning from Hertz he was wrong to do so. But Hertz was right about representation. In physics and much other interesting conversation we do make representations – pictures in words, if you like. In physics we do this by elaborate systems of modelling, structuring, theorizing, calculating, approximating. These are real, articulated, representations of how the world is. The representations of physics are entirely different from simple, non-representational assertions about the location of my typewriter. There is a truth of the matter about the typewriter. In physics there is no final truth of the matter, only a barrage of more or less instructive representations.

Here I have merely repeated at length one of the aphorisms of the turn-of-the-century Swiss-Italian ascetic, Danilo Domodosala: 'When there is a final truth of the matter, then what we say is brief, and it is either true or false. It is not a matter of representation. When, as in physics, we provide representations of the world, there is no final truth of the matter.' Absence of final truth in physics should be the very opposite of disturbing. A correct picture of lively inquiry is given by Hegel, in his preface to the *Phenomenology of Spirit*: 'The True is thus the Bacchanalian revel in which no member is not drunk; yet because each member collapses as he drops out, the revel is just as much transparent and simple repose.' Realism and anti-realism scurry about, trying to latch on to something in the nature of representation that will vanquish the other. There is nothing there. That is why I turn from representing to intervening.

Doing

In a spirit of cheerful irony, let me introduce the experimental part of this book by quoting the most theory-oriented philosopher of recent times, namely Karl Popper:

I suppose that the most central usage of the term 'real' is its use to characterize material things of ordinary size – things which a baby can handle and (preferably) put into his mouth. From this, the usage of the term 'real' is extended, first, to bigger things – things which are too big for us to handle, like railway trains, houses, mountains, the earth and the stars, and also to smaller things – things like dust particles or mites. It is further extended, of course, to liquids and then also to air, to gases and to molecules and atoms.

What is the principle behind the extension? It is, I suggest, that the entities which we conjecture to be real should be able to exert a causal effect upon the *prima facie* real things; that is, upon material things of an ordinary size: that we can explain changes in the ordinary material world of things by the causal effects of entities conjectured to be real.[2]

That is Karl Popper's characterization of our usage of the word 'real'. Note the traditional Lockeian fantasy beginnings. 'Real' is a concept we get from what we, as infants, could put in our mouths. That is a charming picture, not free from nuance. Its absurdity equals that of my own preposterous story of reals and representations. Yet Popper points in the right direction. Reality has to do with causation and our notions of reality are formed from our abilities to change the world.

Maybe there are two quite distinct mythical origins of the idea of 'reality'. One is the reality of representation, the other, the idea of what affects us and what we can affect. Scientific realism is commonly discussed under the heading of representation. Let us now discuss it under the heading of intervention. My conclusion is obvious, even trifling. We shall count as real what we can use to intervene in the world to affect something else, or what the world can use to affect us. Reality as intervention does not even begin to mesh with reality as representation until modern science. Natural science since the seventeenth century has been the adventure of the interlocking of representing and intervening. It is time that philosophy caught up to three centuries of our own past.

2 Karl Popper and John Eccles, *The Self and its Brain*, Berlin, New York and London, 1977, p. 9.

PART B
INTERVENING

9 Experiment

Philosophers of science constantly discuss theories and representation of reality, but say almost nothing about experiment, technology, or the use of knowledge to alter the world. This is odd,
because 'experimental method' used to be just another name for
scientific method. The popular, ignorant, image of the scientist was
someone in a white coat in a laboratory. Of course science preceded
laboratories. Aristotelians downplayed experiment and favoured
deduction from first principles. But the scientific revolution of the
seventeenth century changed all that forever. Experiment was
officially declared to be the royal road to knowledge, and the
schoolmen were scorned because they argued from books instead of
observing the world around them. The philosopher of this revolutionary time was Francis Bacon (1561–1626). He taught that
not only must we observe nature in the raw, but that we must also
'twist the lion's tail', that is, manipulate our world in order to learn
its secrets.

The revolution in science brought with it new institutions. One
of the first was the Royal Society of London, founded about 1660. It
served as the model for other national academies in Paris, St
Petersburg or Berlin. A new form of communication was invented:
the scientific periodical. Yet the early pages of the *Philosophical
Transactions of the Royal Society* have a curious air. Although this
printed record of papers presented to the Society would always
contain some mathematics and theorizing, it was also a chronicle of
facts, observations, experiments, and deductions from experiments. Reports of sea monsters or the weather of the Hebrides rub
shoulders with memorable work by men such as Robert Boyle or
Robert Hooke. Nor would a Boyle or Hooke address the Society
without a demonstration, before the assembled company, of some
new apparatus or experimental phenomenon.

Times have changed. History of the natural sciences is now
almost always written as a history of theory. Philosophy of science

149

has so much become philosophy of theory that the very existence of pre-theoretical observations or experiments has been denied. I hope the following chapters might initiate a Back-to-Bacon movement, in which we attend more seriously to experimental science. Experimentation has a life of its own.

Class and caste

By legend and perhaps by nature philosophers are more accustomed to the armchair than the workbench. It is not so surprising that we have gone overboard for theory at the expense of experiment. Yet we have not always been so insulated. Leibniz has been described as the greatest pure intellect whom the world has ever known. He thought about everything. Although he was less successful in building windmills for mining silver than he was in co-inventing the differential calculus, the remarks of that hyper-intellectual about the role of experiment are undoubtedly more faithful to scientific practice, then or now, than much of what occurs in modern textbooks of philosophy. Philosophers such as Bacon and Liebniz show we don't have to be anti-experimental.

Before thinking about the philosophy of experiments we should record a certain class or caste difference between the theorizer and the experimenter. It has little to do with philosophy. We find prejudices in favour of theory, as far back as there is institutionalized science. Plato and Aristotle frequented the Academy at Athens. That building is located on one side of the Agora, or market place. It is almost as far as possible from the Herculaneum, the temple to the goddess of fire, the patron of the metallurgists. It is 'on the other side of the tracks'. True to this class distinction, we all know a little about Greek geometry and the teachings of the philosophers. Who knows anything about Greek metallurgy? Yet perhaps the gods speak to us in their own way. Of all the buildings that once graced the Athenian Agora, only one stands as it always was, untouched by time or reconstruction. That is the temple of the metallurgists. The Academy fell down long ago. It has been rebuilt – partly by money earned in the steel mills of Pittsburgh.

Even the new science that dedicated itself to experiment maintained a practical bias in favour of the theoretician. I am sure, for example, that Robert Boyle (1627–91) is a more familiar scientific figure than Robert Hooke (1635–1703). Hooke, the experimenter who also theorized, is almost forgotten, while Boyle, the theoretician who also experimented, is still mentioned in primary school text books.

Boyle had a speculative vision of the world as made up of little bouncy or spring-like balls. He was the spokesman for the corpuscular and mechanical philosophy, as it was then called. His important chemical experiments are less well remembered, while Hooke has the reputation of being a mere experimenter – whose theoretical insights are largely ignored. Hooke was the curator of experiments for the Royal Society, and a crusty old character who picked fights with people – partly because of his own lower status as an experimenter. Yet he certainly deserves a place in the pantheon of science. He built the apparatus with which Boyle experimentally investigated the expansion of air (Boyle's law). He discovered the laws of elasticity, which he put to work for example in making spiral springs for pocket watches (Hooke's law). His model of springs between atoms was taken over by Newton. He was the first to build a radical new reflecting telescope, with which he discovered major new stars. He realized that the planet Jupiter rotates on its axis, a novel idea. His microscopic work was of the highest rank, and to him we owe the very word 'cell'. His work on microscopic fossils made him an early proponent of an evolutionary theory. He saw how to use a pendulum to measure the force of gravity. He co-discovered the diffraction of light (it bends around sharp corners, so that shadows are always blurred. More importantly it separates in shadows into bands of dark and light.) He used this as the basis for a wave theory of light. He stated an inverse square law of gravitation, arguably before Newton, although in less perfect a form. The list goes on. This man taught us much about the world in which we live. It is part of the bias for theory over experiment that he is by now unknown to all but a few specialists. It is also due to the fact that Boyle was noble while Hooke was poor and self-taught. The theory/experiment status difference is modelled on social rank.

Nor is such bias a thing of the past. My colleague C.W.F. Everitt wrote on two brothers for the *Dictionary of Scientific Biography*. Both made fundamental contributions to our understanding of superconductivity. Fritz London (1900–53) was a distinguished theoretical low-temperature physicist. Heinz London (1907–70) was a low-temperature experimentalist who also contributed to theory. They were a great team. The biography of Fritz was welcomed by the *Dictionary*, but that of Heinz was sent back for abridgement. The editor (in this case Kuhn) displayed the standard preference for hearing about theory rather than experiment.

Induction and deduction

What is scientific method? Is it the experimental method? The question is wrongly posed. Why should there be *the* method of science? There is not just one way to build a house, or even to grow tomatoes. We should not expect something as motley as the growth of knowledge to be strapped to one methodology.

Let us start with two methodologies. They appear to assign completely different roles to experiment. As examples I take two statements, each made by a great chemist of the last century. The division between them has not expired: it is precisely what separates Carnap and Popper. As I say in the Introduction, Carnap tried to develop a logic of induction, while Popper insists that there is no reasoning except deduction. Here is my own favourite statement of the inductive method:

The foundations of chemical philosophy, are observation, experiment, and analogy. By observation, facts are distinctly and minutely impressed on the mind. By analogy, similar facts are connected. By experiment, new facts are discovered; and, in the progression of knowledge, observation, guided by analogy, leads to experiment, and analogy confirmed by experiment, becomes scientific truth.

To give an instance. – Whoever will consider with attention the slender green vegetable filaments (*Conferva rivularis*) which in the summer exist in almost all streams, lakes, or pools, under the different circumstances of shade and sunshine, will discover globules of air upon the filaments that are shaded. He will find that the effect is owing to the presence of light. This is an *observation*; but it gives no information respecting the nature of the air. Let a wine glass filled with water be inverted over the Conferva, the air will collect in the upper part of the glass, and when the glass is filled with air, it may be closed by the hand, placed in its usual position, and an inflamed taper introduced into it; the taper will burn with more brilliancy than in the atmosphere. This is an *experiment*. If the phenomena are reasoned upon, and the question is put, whether all vegetables of this kind, in fresh or in salt water, do not produce such air under like circumstances, the enquirer is guided by *analogy*: and when this is determined to be the case by new trials, a *general scientific truth* is established – That all Confervae in the sunshine produce a species of air that supports flame in a superior degree; which has been shown to be the case by various minute investigations.

Those are the words with which Humphry Davy (1778–1829) starts his chemistry textbook, *Elements of Chemical Philosophy* (1812, pp. 2–3). He was one of the ablest chemists of his day, commonly remembered for his invention of the miner's safety lamp that prevented many a cruel death, but whose contribution to knowledge includes electrolytic chemical analysis, a technique that enabled him to determine which substances are elements (e.g. chlorine) while others are compounds. Not every chemist shared Davy's inductive view of science. Here are the words of Justus von Liebig (1803–73), the great pioneer of organic chemistry who indirectly revolutionized agriculture by pioneering artificial nitrogen fertilizers.

In all investigations Bacon attaches a great deal of value to experiments. But he understands their meaning not at all. He thinks they are a sort of mechanism which once put in motion will bring about a result of their own. But in science all investigation is deductive or *a priori*. Experiment is only an aid to thought, like a calculation: the thought must always and necessarily precede it if it is to have any meaning. An empirical mode of research, in the usual sense of the term, does not exist. An experiment not preceded by theory, i.e. by an idea, bears the same relation to scientific research as a child's rattle does to music (*Über Francis Bacon von Verulam und die Methode der Naturforschung*, 1863, p. 49).

How deep is the opposition between my two quotations? Liebig says an experiment must be preceded by a theory, that is, an idea. But this statement is ambiguous. It has a weak and a strong version. The *weak version* says only that you must have some ideas about nature and your apparatus before you conduct an experiment. A completely mindless tampering with nature, with no understanding or ability to interpret the result, would teach almost nothing. No one disputes this weak version. Davy certainly has an idea when he experiments on algae. He suspects that the bubbles of gas above the green filaments are of some specific kind. A first question to ask is whether the gas supports burning, or extinguishes it. He finds that the taper flares (from which he infers that the gas is unusually rich in oxygen?) Without that much understanding the experiment would not make sense. The flaring of the taper would at best be a meaningless observation. More likely, no one would even notice. Experiments without ideas like these are not experiments at all.

There is however a *strong version* of Liebig's statement. It says that your experiment is significant only if you are testing a theory about the phenomena under scrutiny. Only if, for example, Davy had the view that the taper would go out (or that it would flare) is his experiment worth anything. I believe this to be simply false. One can conduct an experiment simply out of curiosity to see what will happen. Naturally many of our experiments are made with more specific conjectures in mind. Thus Davy asks whether all algae of the same kind, whether in fresh water or salt, produce gas of this kind, which he doubtless also guesses is oxygen. He makes new trials which lead him to a 'general scientific truth'.

I am not here concerned with whether Davy is really making an inductive inference, as Carnap might have said, or whether he is in the end implicitly following Popper's methodology of conjecture and refutation. It is beside the point that Davy's own example is not, as he thought, a scientific truth. Our post-Davy reclassification of algae shows that *Confervae* are not even a natural kind! There is no such genus or species.

I am concerned solely with the question of the strong version: must there be a conjecture under test in order for an experiment to make sense? I think not. Indeed even the weak version is not beyond doubt. The physicist George Darwin used to say that every once in a while one should do a completely crazy experiment, like blowing the trumpet to the tulips every morning for a month. Probably nothing will happen, but if something did happen, that would be a stupendous discovery.

Which comes first, theory or experiment?

We should not underestimate the generation gap between Davy and Liebig. Maybe the relationship between chemical theory and chemical experiment had changed in the 50 years that separates the two quotations. When Davy wrote, the atomic theory of Dalton and others had only just been stated, and the use of hypothetical models of chemical structures was only just beginning. By the time of Liebig one could no longer practise chemistry by electrically decomposing compounds or identifying gases by seeing whether they support combustion. Only a mind fuelled by a theoretical model could begin to solve mysteries of organic chemicals. We shall find that the relationships between theory and experiment differ at different stages of development, nor do all the natural sciences go through the same cycles. So much may, on reflection, seem all too obvious, but it has been often enough denied, for example by Karl Popper. Naturally we shall expect Popper to be one of, the most forthright of those who prefer theory over experiment. Here is what he does say in his *Logic of Scientific Discovery*:

The theoretician puts certain definite questions to the experimenter, and the latter by his experiments tries to elicit a decisive answer to these questions, and to no others. All other questions he tries hard to exclude. . . . It is a mistake to suppose that the experimenter [. . . aims] 'to lighten the task of the theoretician', or . . . to furnish the theoretician with a basis for inductive generalizations. On the contrary the theoretician must long before have done his work, or at least the most important part of his work: he must have formulated his questions as sharply as possible. Thus it is he who shows the experimenter the way. But even the experimenter is not in the main engaged in making exact observations; his work is largely of a theoretical kind. Theory dominates the experimental work from its initial planning up to the finishing touches in the laboratory (p. 107).

That was Popper's view in the 1934 edition of his book. In the much expanded 1959 edition he adds, in a footnote, that he should have also emphasized, 'the view that observations, and even more so observation statements, and statements of experimental results, are always *interpretations* of the facts observed; that they are *interpretations in the light of theories*'. In a brief initial survey of different relations between theory and experiment, we would do well to start with the obvious counterexamples to Popper. Davy's noticing the bubble of air over the algae is one of these. It was not an 'interpretation in the light of theory' for Davy had initially no theory. Nor was seeing the taper flare an interpretation. Perhaps if he went on to say, 'Ah, then it is oxygen', he would have been making an interpretation. He did not do that.

Noteworthy observations (E)

Much of the early development of optics, between 1600 and 1800, depended on simply noticing some surprising phenomenon. Perhaps the most fruitful of all is the discovery of double refraction in Iceland Spar or calcite. Erasmus Bartholin (1625–98) examined some beautiful crystals brought back from Iceland. If you were to place one of these crystals on this printed page, you would see the print double. Everybody knew about ordinary refraction, and by 1689, when Bartholin made his discovery, the laws of refraction were well known, and spectacles, the microscope and the telescope were familiar. This background makes Iceland Spar remarkable at two levels. Today one is still surprised and delighted by these crystals. Moreover there was a surprise to the physicist of the day, knowing the laws of refraction, who notes that in addition to the ordinary refracted ray there is an 'extraordinary' one, as it is still called.

Iceland Spar plays a fundamental role in the history of optics, because it was the first known producer of polarized light. The phenomenon was understood in a very loose way by Huygens, who proposed that the extraordinary ray had an elliptical, rather than a spherical, wave surface. However our present understanding had to wait until the wave theory of light was revived. Fresnel (1788–1827), the founder of modern wave theory, gave a magnificent analysis in which the two rays are described by a single equation whose solution is a two-sheeted surface of the fourth degree. Polarization has turned out, time and again, to lead us ever deeper into the theoretical understanding of light.

There is a whole series of such 'surprising' observations. Grimaldi (1613–63) and then Hooke carefully examined something of which we are all vaguely aware – that there is some illumination in the shadow of an opaque body. Careful observation revealed regularly spaced bands at the edge of the shadow. This is called diffraction, which originally meant 'breaking into pieces' of the light in these bands. These observations preceded theory in a characteristic way. So too did Newton's observation of the dispersion of light, and the work by Hooke and Newton on the colours of thin plates. In due course this led to interference phenomena called Newton's rings. The first quantitative explanation of this phenomenon was not made until more than a century later, in 1802, by Thomas Young (1773–1829).

Now of course Bartholin, Grimaldi, Hooke and Newton were not mindless empiricists without an 'idea' in their heads. They saw what they saw because they were curious, inquisitive, reflective people. They were attempting to form theories. But in all these cases it is clear that the observations preceded any formulation of theory.

The stimulation of theory (E)

At a later epoch we find similar noteworthy observations that stimulate theory. For example in 1808 polarization by reflection was discovered. A colonel in Napoleon's corps of engineers, E.L. Malus (1775–1812), was experimenting with Iceland Spar and noticed the effects of evening sunlight being reflected from the windows of the nearby Palais du Luxembourg. The light went through his crystal when it was held in a vertical plane, but was blocked when the crystal was held in a horizontal plane. Similarly, fluorescence was first noticed by John Herschel (1792–1871) in 1845, when he began to pay attention to the blue light emitted in a solution of quinine sulfate when it was illuminated in certain ways.

Noteworthy observation must, of its nature, be only the beginning. Might one not grant the point that there are initial observations that precede theory, yet contend that all deliberate experimentation is dominated by theory, just as Popper says? I think not. Consider David Brewster (1781–1868), a by now forgotten but once prolific experimenter. Brewster was the major figure in experimental optics between 1810 and 1840. He determined the laws of reflection and refraction for polarized light. He was able to induce birefringence (i.e. polarizing properties) in bodies under stress. He discovered biaxial double refraction and made the first and fundamental steps towards the complex laws of metallic reflection. We now speak of Fresnel's laws, the sine and tangent laws for the intensity of reflected polarized light, but Brewster published them in 1818, five years before Fresnel's treatment of them within wave theory. Brewster's work established the material on which many developments in the wave theory were to be based. Yet in so far as he had any theoretical views, he was a dyed in the wool Newtonian, believing light consists of rays of corpuscles. Brewster was not testing or comparing theories at all. He was trying to find out how light behaves.

Brewster firmly held to the 'wrong' theory while creating the experimental phenomena that we can understand only with the 'right' theory, the very theory that he vociferously rejected. He did not 'interpret' his experimental findings in the light of his wrong theory. He made some phenomena for which any theory must, in the end, account. Nor is Brewster alone in this. A more recent brilliant experimenter was R.W. Wood (1868–1955) who between 1900 and 1930 made fundamental contributions to quantum optics, while remaining almost entirely innocent of, and sceptical about, quantum mechanics. Resonance radiation, fluorescence, absorption spectra, Raman spectra – all these require a quantum mechanical understanding, but Wood's contribution arose not from the theory but, like Brewster's, from a keen ability to get nature to behave in new ways.

Meaningless phenomena

I do not contend that noteworthy observations in themselves do anything. Plenty of phenomena attract great excitement but then have to lie fallow because no one can see what they mean, how they connect with anything else, or how they can be put to some use. In 1827 a botanist, Robert Brown, reported on the irregular movement of pollen suspended in water. This Brownian motion had been observed by others even 60 years before; some thought it was vital action of living pollen itself. Brown made painstaking observations, but for long it came to nothing. Only in the first decade of the present century did we have simultaneous work by experimenters, such as J. Perrin, and theoreticians, such as Einstein, which showed that the pollen was being bounced around by molecules. These results were what finally converted even the greatest sceptics to the kinetic theory of gases.

A similar story is to be told for the photoelectric effect. In 1839 A.-C. Becquerel noticed a very curious thing. He had a small electrovoltaic cell, that is, a pair of metal plates immersed in a dilute acid solution. Shining a light on one of the plates changed the voltage of the cell. This attracted great interest – for about two years. Other isolated phenomena were noticed. Thus the resistance of the metal selenium was decreased simply by illuminating it (1873). Once again it was left to Einstein to figure out what was happening; to this we owe the theory of the photon and innumerable familiar applications, including television (photoelectric cells convert the light reflected from an object into electric currents).

Thus I make no claim that experimental work could exist independently of theory. That would be the blind work of those whom Bacon mocked as 'mere empirics'. It remains the case, however, that much truly fundamental research precedes any relevant theory whatsoever.

Happy meetings

Some profound experimental work is generated entirely by theory. Some great theories spring from pre-theoretical experiment. Some theories languish for lack of mesh with the real world, while some experimental phenomena sit idle for lack of theory. There are also happy families, in which theory and experiment coming from different directions meet. I shall give an example in which sheer dedication to an experimental freak led to a firm fact which suddenly meshed with theories coming from an entirely different quarter.

In the early days of transatlantic radio there was always a lot of static. Many sources of the noise could be identified, although they could not always be removed. Some came from electric storms. Even in the 1930s Karl Jansky at the Bell Telephone Laboratories had located a 'hiss' coming from the centre of the Milky Way. Thus there were sources of radio energy in space which contributed to the familiar static.

In 1965 the radioastronomers Arno Penzias and R.W. Wilson adapted a radiotelescope to study this phenomenon. They expected to detect energy sources and that they did. But they were also very diligent. They found a small amount of energy which seemed to be everywhere in space, uniformly distributed. It would be as if everything in space which was not an energy source were about 4°K. Since this did not make much sense, they did their best to discover instrumental errors. For example, they thought that some of this radiation might come from the pigeons that were nesting on their telescope, and they had a dreadful time trying to get rid of the pigeons. But after they had eliminated every possible source of noise, they were left with a uniform temperature of 3°K. They were loth to publish because a completely homogeneous background radiation did not make much sense.

Fortunately, just as they had become certain of this meaningless phenomenon, a theoretical group, at Princeton, was circulating a preprint which suggested, in a qualitative way, that if the universe had originated in a Big Bang, there would be a uniform temperature throughout space, the residual temperature of the first explosion. Moreover this energy would be detected in the form of radio signals. The experimental work of Penzias and Wilson meshed beautifully with what would otherwise have been mere speculation.

Penzias and Wilson had showed that the temperature of the universe is almost everywhere about three degrees above absolute zero; this is the residual energy of creation. It was the first truly compelling reason to believe in that Big Bang.

It is sometimes said that in astronomy we do not experiment; we can only observe. It is true that we cannot interfere very much in the distant reaches of space, but the skills employed by Penzias and Wilson were identical to those used by laboratory experimenters. Shall we say with Popper, in the light of this story, that in general 'the theoretician must long before have done his work, or at least the most important part of his work: he must have formulated his questions as sharply as possible. Thus it is he who shows the experimenter the way'? Or shall we say that although some theory precedes some experiment, some experiment and some observation precedes theory, and may for long have a life of its own? The happy family I have just described is the intersection of theory and skilled observation. Penzias and Wilson are among the few experimenters in physics to have been given a Nobel Prize. They did not get it for refuting anything, but for exploring the universe.

Theory-history

It may seem that I have been overstating the way that theory-dominated history and philosophy of science skew our perception of experiment. In fact it is understated. For example, I have related the story of three degrees just as it is told by Penzias and Wilson themselves, in their autobiographical film *Three Degrees*.[1] They were exploring, and found the uniform background radiation prior to any theory of it. But here is what happens to this very experiment when it becomes 'history':

Theoretical astronomers have predicted that if there had been an explosion billions of years ago, cooling would have been going on ever since the event. The amount of cooling would have reduced the original temperature of perhaps a billion degrees to $3°K - 3°$ above absolute zero.

Radioastronomers believed that if they could aim a very sensitive receiver at a blank part of the sky, a region that appeared to be empty, it might be possible to determine whether or not the theorists were correct. This was done in the early 1970s. Two scientists at Bell Telephone Laboratories (the same place where Karl Jansky had discovered cosmic radio waves) picked up radio

1 Information and Publication Division, Bell Laboratories, 1979.

signals from 'empty' space. After sorting out all known causes for the signals, there was still left a signal of 3° they could not account for. Since that first experiment others have been carried out. They always produce the same result – 3° radiation.

Space is not absolutely cold. The temperature of the universe appears to be 3°K. It is the exact temperature the universe should be if it all began some 13 billion years ago, with a Big Bang.[2]

We have seen another example of such rewriting of history in the case of the muon or meson, described in Chapter 6. Two groups of workers detected the muon on the basis of cloud chamber studies of cosmic rays, together with the Bethe–Heitler energy-loss formula. History now has it that they were actually looking for Yukawa's 'meson', and mistakenly thought they had found it – when in fact they had never heard of Yukawa's conjecture. I do not mean to imply that a competent historian of science would get things so wrong, but rather to notice the constant drift of popular history and folklore.

Ampère, theoretician

Let it not be thought that, in a new science, experiment and observation precede theory, even if, later on, theory will precede observation. A.-M. Ampère (1775–1836) is a fine example of a great scientist starting out on a theoretical footing. He had primarily worked in chemistry, and produced complex models of atoms which he used to explain and develop experimental investigations. He was not especially successful at this, although he was one of those who, independently, about 1815, realized what we now call Avogadro's law, that equal volumes of gases at equal temperature and pressure will contain exactly the same number of molecules, regardless of the kind of gas. As we have already seen in Chapter 7 above, he much admired Kant, and insisted that theoretical science was a study of noumena behind the phenomena. We form theories about the things in themselves, the noumena, and are thereby able to explain the phenomena. That was not exactly what Kant intended, but no matter. Ampère was a theoretician whose moment came on September 11 1820. He saw a demonstration by Ørsted that a compass needle is deflected by an electric current. Commencing on September 20 Ampère laid out, in weekly lectures, the

2 F.M. Bradley, *The Electromagnetic Spectrum*, New York, 1979, p. 100, my emphasis.

foundations of the theory of electromagnetism. He made it up as he went along.

That, at any rate, is the story. C.W.F. Everitt points out that there must be more to it than that, and that Ampère, having no official post-Kantian methodology of his own, wrote his work to fit. The great theoretician–experimenter of electromagnetism, James Clerk Maxwell, wrote a comparison of Ampère and Humphry Davy's pupil Michael Faraday, praising both 'inductivist' Faraday and 'deductivist' Ampère. He described Ampère's investigation as 'one of the most brilliant achievements in science . . . perfect in form, unassailable in accuracy . . . summed up in a formula from which all the phenomena may be deduced', but then went on to say that whereas Faraday's papers candidly reveal the workings of his mind,

We can scarcely believe that Ampère really discovered the law of action: by means of the experiments which he describes. We are led to suspect what, indeed, he tells us himself, that he discovered the law by some process he has not shewn us, and that when he had afterwards built up a perfect demonstration he removed all traces of the scaffolding by which he had raised it.

Mary Hesse remarks, in her *Structure of Scientific Inference* (pp. 201f, 262), that Maxwell called Ampère the Newton of electricity. This alludes to an alternative tradition about the nature of induction, which goes back to Newton. He spoke of deduction from phenomena, which was an inductive process. From the phenomena we infer propositions that describe them in a general way, and then are able, upon reflection, to create new phenomena hitherto unthought of. That, at any rate, was Ampère's procedure. He would usually begin one of his weekly lectures with a phenomenon, demonstrated before the audience. Often the experiment that created the phenomenon had not existed at the end of the lecture of the preceding week.

Invention (E)

A question posed in terms of theory and experiment is misleading because it treats theory as one rather uniform kind of thing and experiment as another. I turn to the varieties of theory in Chapter 12. We have seen some varieties in experiment, but there are also other relevant categories, of which invention is one of the most important. The history of thermodynamics is a history of practical invention that gradually leads to theoretical analysis. One road to new technology is the elaboration of theory and experiment which is then applied to practical problems. But there is another road, in which the inventions proceed at their own practical pace and theory spins off on the side. The most obvious example is the best one: the steam engine.

There were three phases of invention and several experimental concepts. The inventions are Newcomen's atmospheric engine (1709–15), Watt's condensing engine (1767–84) and Trevithick's high-pressure engine (1798). Underlying half the developments after Newcomen's original invention was the concept, as much one of economics as of physics, of the 'duty' of an engine, that is, the number of foot-pounds of water pumped per bushel of coal. Who had the idea is not known. Probably it was not anyone recorded in a history of science but rather the hard-headed value-for-money outlook of the Cornish mine-managers, who noticed that some engines pumped more efficiently than others and did not see why they should be short-changed when the neighbouring mine had a better rating. At first, the success of Newcomen's engine hung in the balance because, except in deep mines, it was only marginally cheaper to operate than horse-driven pumps. Watt's achievement, after seventeen years of trial and error, was to produce an engine guaranteed to have a duty at least four times better than the best Newcomen engine. (Imagine a marketable motor car with the same power as existing cars but capable of doing 100 miles per gallon instead of 25.)

Watt first introduced the separate condenser, then made the engine double-acting, that is, let in steam on one side of the cylinder while pulling a vacuum on the other, and finally in 1782 introduced the principle of expansive working, that is, cutting off the flow of steam into the cylinder early in its stroke, and allowing it to expand the rest of the way under its own pressure. Expansive working means some loss of power from an engine of a given size, but an increase in 'duty'. Of these ideas, the most important for pure science was expansive working. A very useful practical aid, devised about 1790 by Watt's associate, James Southern, was the *indicator diagram*. The indicator was an automatic recorder which could be attached to the engine to plot pressure in the cylinder against the volume measured from the stroke: the area of the curve so traced was a measure of the work done in each stroke. The indicator was used to tune the engine to maximum performance. That very diagram became part of the Carnot cycle of theoretical thermodynamics.

Trevithick's great contribution, at first more a matter of courage than of theory, was to go ahead with building a high-pressure engine despite the danger of explosions. The first argument for high-pressure working is compactness: one can get more power from an engine of a given size. So Trevithick built the first successful locomotive engine in 1799. Soon another result emerged. If the high-pressure engine was worked expansively with early cut-off, its duty became higher (ultimately much higher) than the best Watt engine. It required the genius of Sadi Carnot (1796–1832) to come to grips with this phenomenon and see that the advantage of the high-pressure engine is not pressure alone, but the increase in the boiling point of water with pressure. The efficiency of the engine depends not on pressure differences but on the temperature difference between the steam entering the cylinder and the expanded steam leaving the cylinder. So was born the Carnot cycle, the concept of thermodynamic efficiency, and finally when Carnot's ideas had been unified with the principle of conservation of energy, the science of thermodynamics.

What indeed does 'thermodynamics' mean? The subject deals not with the flow of heat, which might be called its dynamics, but with what might be called thermostatic phenomena. Is it misnamed? No. Kelvin coined the words 'thermo-dynamic engine' in 1850 to describe any machine like the steam engine or Carnot's ideal engine. These engines were called dynamic because they convert heat into work. Thus the very word 'thermodynamics' recalls that this science arose from a profound analysis of a notable sequence of inventions. The development of that technology involved endless 'experiment' but not in the sense of Popperian testing of theory nor of Davy-like induction. The experiments were the imaginative trials required for the perfection of the technology that lies at the centre of the industrial revolution.

A multitude of experimental laws, waiting for a theory (E)

The *Theory of the Properties of Metals and Alloys* (1936) is a standard old textbook whose distinguished authors, N.F. Mott and H. Jones, discuss, among other things, the conduction of electricity and heat in various metallic substances. What must a decent theory of this subject cover? Mott and Jones say that a theory of metallic conduction has to explain, among others, the following experimental results:

(1) The Wiedemann–Franz law which states that the ratio of the thermal to the electrical conductivity is equal to LT, where T is the absolute temperature and L is a constant which is the same for all metals.

(2) The absolute magnitude of the electrical conductivity of a pure metal, and its dependence on the place of the metal in the periodic table, e.g., the large conductivities of the monovalent metals and the small conductivities of the transition metals.

(3) The relatively large increases in the resistance due to small amounts of impurities in solid solution, and the Matthiessen rule, which states that the change in resistance due to a small quantity of foreign metal in solid solution is independent of the temperature.

(4) The dependence of the resistance on temperature and on pressure.

(5) The appearance of supraconductivity [superconductivity].

Mott and Jones go on to say that 'with the exception of (5) the theory of conductivity based on quantum mechanics has given at least a qualitative understanding of all these results' (p. 27). (A quantum mechanical understanding of superconductivity was eventually reached in 1957.)

The experimental results in this list were established long before there was a theory around to fit them together. The Wiedemann–Franz law (1) dates from 1853, Matthiessen's rule from 1862 (3), the relationships between conductivity and position in the periodic table from the 1890s (2), and superconductivity (5) from 1911. The data were all there; what was needed was a coordinating theory. The difference between this case and that of optics and thermodynamics is that the theory did not come directly out of the data, but from much more general insights into atomic structure. Quantum mechanics was both the stimulus and the solution. No one could sensibly suggest that the organization of the phenomenological laws within the general theory is a mere matter of induction, analogy or generalization. Theory has in the end been crucial to knowledge, to the growth of knowledge, and to its applications. Having said that, let us not pretend that the various phenomenological laws of solid state physics required a theory – any theory – before they were known. Experimentation has many lives of its own.

Too many instances?

After this Baconian fluster of examples of many different relationships between experiment and theory, it may seem as if no
statements of any generality are to be made. That is already an
achievement, because, as the quotations from Davy and Liebig
show, any one-sided view of experiment is certainly wrong. Let us
now proceed to some positive ends. What is an observation? Do we
see reality through a microscope? Are there crucial experiments?
Why do people measure obsessively a few quantities whose value, at
least to three places of decimals, is of no intrinsic interest to theory
or technology? Is there something in the nature of experimentation
that makes experimenters into scientific realists? Let us begin at the
beginning. What is an observation? Is every observation in science
loaded with theory?

10 Observation

Commonplace facts about observation have been distorted by two philosophical fashions. One is the vogue for what Quine calls semantic ascent (don't talk about things, talk about the way we talk about things). The other is the domination of experiment by theory. The former says not to think about observation, but about observation statements – the words used to report observations. The latter says that every observation statement is loaded with theory – there is no observing prior to theorizing. Hence it is well to begin with a few untheoretical unlinguistic platitudes.

1 Observation, as a primary source of data, has always been a part of natural science, but it is not all that important. Here I refer to the philosophers' conception of observation: the notion that the life of the experimenter is spent in the making of observations which provide the data that test theory, or upon which theory is built. This kind of observation plays a relatively minor role in most experiments. Some great experimenters have been poor observers. Often the experimental task, and the test of ingenuity or even greatness, is less to observe and report, than to get some bit of equipment to exhibit phenomena in a reliable way.

2 There is, however, a more important and less noticed kind of observation that is essential to fine experimentation. The good experimenter is often the observant one who sees the instructive quirks or unexpected outcomes of this or that bit of the equipment. You will not get the apparatus working unless you are observant. Sometimes persistent attention to an oddity that would have been dismissed by a lesser experimenter is precisely what leads to new knowledge. But this is less a matter of the philosophers' observation-as-reporting-what-one-sees, than the sense of the word we use when we call one person observant while another is not.

3 Noteworthy observations, such as those described in the previous chapter, have sometimes been essential to initiating

inquiry, but they seldom dominate later work. Experiment super-
sedes raw observation.

4 Observation is a skill. Some people are better at it than others.
You can often improve this skill by training and practise.

5 There are numerous distinctions between observation and
theory. The philosophical idea of a pure 'observation statement'
has been criticized on the ground that all statements are theory-
loaded. This is the wrong ground for attack. There are plenty of
pre-theoretical observation statements, but they seldom occur in
the annals of science.

6 Although there is a concept of 'seeing with the naked eye',
scientists seldom restrict observation to that. We usually observe
objects or events with instruments. The things that are 'seen' in
twentieth-century science can seldom be observed by the unaided
human senses.

Observation has been over-rated

Much of the discussion about observation, observation statements
and observability is due to our positivist heritage. Before positivism, observation is not central. Francis Bacon is our early philosopher of the inductive sciences. You might expect him to say a lot
about observations. In fact he appears not even to use the word.
Positivism had not yet struck.

The word 'observation' was current in English when Bacon
wrote, and applied chiefly to observations of the altitude of heavenly
bodies, such as the sun. Hence from the very beginning, observation was associated with the use of instruments. Bacon uses a more
general term of art, often translated by the curious phrase,
prerogative instances. In 1620 he listed 27 different kinds of these.
Included are what we now call crucial experiments, which he called
crucial instances, or more correctly, instances of the crossroads
(*instantiae crucis*). Some of Bacon's 27 kinds of instances are pre-
theoretical noteworthy observations. Others are motivated by a
desire to test theory. Some are made with devices that 'aid the
immediate actions of the senses'. These include not only the new
microscopes and Galileo's telescope but also 'rods, astrolabes and
the like; which do not enlarge the sense of sight, but rectify and
direct it'. Bacon moves on to 'evoking' devices that 'reduce the
non-sensible to the sensible; that is, make manifest, things not

directly perceptible, by means of others which are'. (*Novum Organum* Secs. xxi–lii.)

Bacon thus knows the difference between what is directly perceptible and those invisible events which can only be 'evoked'. The distinction is, for Bacon, both obvious and unimportant. There is some evidence that it really matters only after 1800, when the very conception of 'seeing' undergoes something of a transformation. After 1800, to see is to see the opaque surface of things, and all knowledge must be derived from this avenue. This is the starting point for both positivism and phenomenology. Only the former concerns us here. To positivism we owe the need to distinguish sharply between inference and seeing with the naked eye (or other unaided senses).

Positivist observation

The positivist, we recall, is against causes, against explanations, against theoretical entities and against metaphysics. The real is restricted to the observable. With a firm grip on observable reality the positivist can do what he wants with the rest.

What he wants for the rest varies from case to case. The logical positivists liked the idea of using logic to 'reduce' theoretical statements, so that theory becomes a logical short-hand for expressing facts and organizing thoughts about what can be observed. On one version this would lead to a wishy-washy scientific realism: theories may be true, and the entities that they mention may exist, so long as none of that talk is understood too literally.

In another version of logical reduction, the terms referring to theoretical entities would be shown, on an analysis, not to have the logical structure of referring terms at all. Since they are not referential, they don't refer to anything, and theoretical entities are not real. This use of reduction leads to a fairly stringent anti-realism. But since nobody has made a logical reduction of any interesting natural science, such questions are vacuous.

The positivist then takes another tack. He may say with Comte or van Fraassen that theoretical statements are to be understood literally, but not to be believed. As the latter puts it, in *The Scientific Image*, 'When a scientist advances a new theory, the realist sees him as asserting the (truth of the) postulate. But the anti-realist sees him as displaying this theory, holding it up to view, as it were, and claiming certain virtues for it' (p. 27). A theory may be accepted because it accounts for phenomena and helps in prediction. It may be accepted for its pragmatic virtues without being believed to be literally true.

Positivists such as Comte, Mach, Carnap or van Fraassen insist in these various ways that there is a distinction between theory and observation. That is how they make the world safe from the ravages of metaphysics.

Denying the distinction

Once the distinction between observation and theory was made so important, it was certain to be denied. There are two grounds of denial. One is conservative, and realist in its tendencies. The other is radical, more romantic, and often leans towards idealism. There was an outburst of both kinds of response around 1960.

Grover Maxwell exemplifies the realist response. In a 1962 paper he says that the contrast between being observable and merely theoretical is vague. It often depends more on technology than on anything in the constitution of the world.[1] Nor, he continues, is the distinction of much importance to natural science. We cannot use it to argue that no theoretical entities really exist.

In particular Maxwell says that there is a continuum that starts with seeing through a vacuum. Next comes seeing through the atmosphere, then seeing through a light microscope. At present this continuum may end with seeing using a scanning electron microscope. Objects like genes which were once merely theoretical are transformed into observable entities. We now see large molecules. Hence observability does not provide a good way to sort the objects of science into real and unreal.

Maxwell's case is not closed. We should attend more closely to the very technologies that he takes for granted. I attempt this in the next chapter, on microscopes. I agree with Maxwell's playing down of visibility as a basis for ontology. In a paper I discuss later in this chapter, Dudley Shapere makes the further point that physicists regularly talk about observing and even seeing using devices in which neither the eye nor any other sense organ could play any

---

1 G. Maxwell, 'The ontological status of theoretical entities', *Minnesota Studies in the Philosophy of Science* 3 (1962), pp. 3–27.

essential role at all. In his example, we try to observe the interior of the sun using neutrinos emitted by solar fusion processes. What counts as an observation, he says, itself depends upon current theory. I shall return to this theme, but first we should look at the more daring and idealist-leaning rejection of the distinction between theory and observation. Maxwell said that the observability of *entities* has nothing to do with their ontological status. Other philosophers, at the same time, were saying that there are no purely observation *statements* because they are all infected by theory. I call this idealist-leaning because it makes the very content of the feeblest scientific utterances determined by how we think, rather than mind-independent reality. We can diagram these differences in the following way:

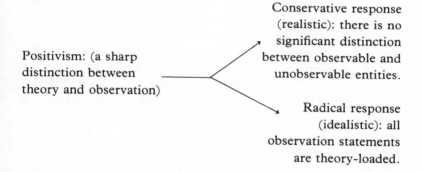

Positivism: (a sharp distinction between theory and observation)

Conservative response (realistic): there is no significant distinction between observable and unobservable entities.

Radical response (idealistic): all observation statements are theory-loaded.

Theory-loaded

In 1959 N.R. Hanson gave us the catchword 'theory-loaded' in his splendid book, *Patterns of Discovery*. The idea is that every observational term and sentence is supposed to carry a load of theory with it.

One fact about language tends to dominate those parts of *Patterns of Discovery* in which the word 'theory-loaded' occurs. We are reminded that there are very subtle linguistic rules about even the most commonplace words, for example the verb 'to wound' and the noun 'wound'. Only some cuts, injuries, etc., in quite specific kinds of situations, count as wounds. If a surgeon describes a gash in a man's leg as a wound, that may imply that the man was hurt in a fight or in battle. Such implications occur all the time, but they are not in my opinion worth calling theoretical assumptions. This part of the theory-loaded doctrine is an important and unexceptionable assertion about ordinary language. It in no way implies that all reports of observation must carry a load of scientific theory.

Hanson also points out that we tend to notice things only when we have expectations, often of a theoretical sort, which will make them seem interesting or at least to make sense. That is true but it is different from the theory-loaded doctrine. I shall turn to it presently. First, I address some more dubious claims.

Lakatos on observation

Lakatos, for example, says that the simplest kind of falsificationism – the kind we often attribute to Popper – won't do because it takes for granted a theory/observation distinction. We cannot have the simple rule about theories, that people propose them and nature disposes of them. That, says Lakatos, rests on two false assumptions. First, that there is a psychological borderline between speculative propositions and observational ones, and, secondly, that observational propositions can be proved by (looking at) the facts. For the past 15 years these assumptions have been jeered at, but we ought also to have argument. Lakatos's arguments are dismayingly facile and ineffective. He says that a 'few characteristic examples already undermine the first assumption'. In fact he gives one example, of Galileo using a telescope to see sun-spots, a seeing which cannot be purely observational. Is that supposed to refute, or even undermine, the theory/observation distinction?

As for the second point, that one can look and see whether observation sentences are true, Lakatos writes in italics, 'no factual proposition can ever be proved from an experiment . . . one cannot prove statements from experience. . . . This is one of the basic points of elementary logic, but one which is understood by relatively few people even today' (I, p. 16). Such an equivocation on the verb 'prove' is particularly disheartening from a writer from whom I learned the several senses of the verb: that the verb properly bears the sense of 'test' (the proof of the pudding is in the eating, galley proofs), and that such tests often lead to establishing facts (the pudding is stodgy, the galleys full of misprints).

On containing theoretical assumptions

Paul Feyerabend's essays, contemporary with work by Hanson, also played down the distinction between theory and observation.

He has since come to dismiss the philosophical obsession with language and meanings. He has denounced the very phrase, 'theory-laden'. But this is not because he thinks that some of what we say is free from theory. Quite the contrary. To call statements theory-laden, he says, is to suggest that there is a sort of observational truck on to which a theoretical component is loaded. There is no such truck. Theory is everywhere.

In his most famous book, *Against Method* (1977), Feyerabend says that there is no point to the distinction between theory and observation. Curiously, for all his avowed rejection of linguistic discussions, he still speaks as if the theory/observation distinction were a distinction between sentences. He suggests it is just a matter of obvious and less obvious sentences, or between long ones and short ones. 'Nobody will deny that such distinctions *can be made*, but nobody will put great weight on them, for they do not play any decisive role in the business of science.' (p. 168). We also read what sounds like the 'theory-loaded' doctrine in full force: 'observational reports, experimental results, "factual statements", either *contain* theoretical assumptions or assert them by the manner in which they are used.' (p. 31). I disagree with what is actually said here, but before explaining why, I want to cancel something suggested by remarks like this. They give the idea that experimental results exhaust what matters to an experiment, and that experimental results are stated by, or even constituted by, an observation report or a 'factual statement'. I shall insist on the truism that experimenting is not stating or reporting but doing – and not doing things with words.

Statements, records, results

Observation and experiment are not one thing nor even opposite poles of a smooth continuum. Evidently many observations of interest have nothing to do with experiments. Claude Bernard's 1865 *Introduction to the Study of Experimental Medicine* is the classic attempt to distinguish the concepts of experiment and observation. He tests his classification by a lot of difficult examples from medicine where observation and experiment get muddled up. Consider Dr Beauchamp who, in the Anglo-American war of 1812, had the good fortune to observe, over an extended period of time, the workings of the digestive tract of a man with a dreadful stomach wound. Was that an experiment or just a sequence of fateful observations in almost unique circumstances? I do not want to pursue such points, but instead to emphasize something that is more noticeable in physics than medicine.

The Michelson–Morley experiment has the merit of being well known. It is famous because with hindsight it seemed to some historians to refute the entire theory of the electromagnetic aether, and thus to be the experimental forerunner of Einstein's theory of relativity. The chief published *report* of the experiment of 1887 is 12 pages long. The *observations* were made in the course of a couple of hours on July 8, 9, 11, and 12. The *results* of the experiment are notoriously controversial; Michelson thought the chief result was a refutation of the earth's motion relative to the aether. As I show in Chapter 15 below, he also thought that it discredited a theory used to explain why the stars are not quite where they appear to be. At any rate the experiment lasted over a year. This included making and remaking the apparatus and getting it to work, and above all acquiring the curious knack of knowing when the apparatus is working. It has been common practice to use the label 'the Michelson–Morley experiment' to denote a sequence of intermittent work with Michelson's initial success of 1881 (or even earlier, some failures) and going on to include Miller's work of the 1920s. One could say that the experiment lasted half a century, while the observations lasted maybe a day and a half. Moreover the chief result of the experiment, although not an experimental result, was a radical transformation in the possibilities of measurement. Michelson won a Nobel prize for this, not for his impact on aether theories.

In short Feyerabend's 'factual statements, observation reports, and experimental results' are not even the same kinds of thing. To lump them together is to make it almost impossible to notice anything about what goes on in experimental science. In particular they have nothing to do with Feyerabend's difference between long and short sentences.

Observation without theory

Feyerabend says that observational reports, etc., always contain or assert theoretical assumptions. This assertion is hardly worth debating because it is obviously false, unless one attaches a quite attenuated sense to the words, in which case the assertion is true but trivial.

Most of the verbal quibble arises over the word 'theory', a word best reserved for some fairly specific body of speculation or propositions with a definite subject matter. Unfortunately the Feyerabend of my quotation used the word 'theory' to denote all sorts of inchoate, implicit, or imputed beliefs. To condense him without malice, he wrote of some alleged habits and beliefs:

Our habit of saying the table is brown when we view it under normal circumstances, or saying the table seems brown when viewed under other circumstances . . . our belief that some of our sensory impressions are veridical and some are not . . . that the medium between us and the object does not distort . . . that the physical entity that establishes the contact carries a true picture. . . .

All these are supposed to be theoretical assumptions underlying our commonplace observations, and 'the material which the scientist has at his disposal, his most sublime theories and his most sophisticated techniques included, is structured in exactly the same way'.

Now taken literally most of this is, to be polite, rather hastily said. For example, what is this 'habit of saying the table is brown when we view it under normal circumstances'? I doubt that ever in my life, before, have I uttered either the sentence 'the table is brown' or the 'table seems to be brown'. I am certainly not in the habit of uttering the first sentence when looking at a table in a good light. I have only met one person with any such habit, a French lunatic who habitually and repeatedly uttered, *C'est de la merde, ça* whenever he saw excrement in normal viewing conditions, for example, when we were manuring a field. Nor would I impute to poor Boul-boul any of the assumptions listed by Feyerabend. Feyerabend has shown us how not to talk about observation, speech, theory, habits, or reporting.

Of course we have all sorts of expectations, prejudices, opinions, working hypotheses and habits when we say anything. Some of these we express. Some are contextual implications. Some can be imputed to the speaker by a sensitive student of the human mind. Some propositions which could be assumptions or presuppositions in another context are not so in the context of routine existence. Thus I could make the assumption that the air between me and the printed page does not distort the shapes of the words I see, and I could *perhaps* investigate this assumption. (How?) But when I read aloud, or make corrections on this page I simply interact with something of interest to me, and it is wrong to speak of assumptions. In particular it is wrong to speak of theoretical assumptions. I have not the remotest idea what a theory of non-distortion by the air would be like. Of course if you want to call every belief, proto-belief, and belief that could be invented, a theory, do so. But then the claim about theory-loaded is trifling.

There have been important observations in the history of science, which have included no theoretical assumptions at all. The noteworthy observations of the previous chapter furnish examples. Here is another, of more recent date, where we can set down a pristine observation statement.

Herschel and radiant heat

William Herschel was an adroit and insatiable searcher of the midnight sky, builder of the greatest telescope of his time and immensely extending our catalogue of the heavens. Here I consider an incidental event of 1800, when Herschel was 61. That was the year in which, as we now put it, he discovered radiant heat. He made about 200 experiments and published four major papers on the topic, of which the last is 100 pages long. All are to be found in the *Philosophical Transactions of the Royal Society* for 1800. He began by making what we now think of as the right proposal about radiant heat, but ended up in a quandary, not sure where the truth might lie.

He had been using coloured filters in one of his telescopes. He noticed that filters of different colours transmit different amounts of heat: 'When I used some of them I felt a sensation of heat, though I had but little light, while others gave me much light with scarce any sensation of heat.' We shall not find a better sense-datum report than this, in the whole of physical science. Naturally we remember it not for its sensory quality but because of what came next. Why did Herschel do anything next? First of all he wanted filters better suited for looking at the sun. Certainly he also had his mind on certain speculative issues that were then coming to the fore.

He used thermometers to study the heating effect of rays of light separated with a prism. This really set him going, for he found not only that orange warms more than indigo, but that there is also a heating effect below the visible red spectrum. His first guess about this phenomenon was roughly what we now believe. He took it that both visible and invisible rays are emitted from the sun. Our eyes are sensitive to only one part of the spectrum of radiation. We are warmed by a different overlapping part. Since he believed in the Newtonian corpuscular theory of light, he thought in terms of rays composed of particles. Sight responds to corpuscles of violet through red, while the sense of heat responds to corpuscles of yellow through infra-red.

He now set out to investigate this idea by seeing whether heat and light rays in the visible spectrum have the same properties. So he compared their reflection, refraction and differential refrangibility, their tendency to be stopped by diaphanous bodies, and their liability to scattering from rough surfaces.

At this stage in Herschel's papers we have a large number of observations of various angles, proportions of light transmitted and so forth. He certainly has an experimental idea, but only one of a rather nebulous sort. His theory was entirely Newtonian: he thought that light consisted of rays of particles, but this had limited impact upon the details of his research. His difficulties were not theoretical but experimental. Photometry – the practice of measuring aspects of transmitted light – had been in fair state for 40 years, but calorimetry was almost nonexistent. There were procedures for filtering out rays of light, but how should one filter rays of heat? Herschel was probing phenomena. He made many claims to accuracy which we now think to be misplaced. He measured not only transmission of light but also transmission of heat to one part in a thousand. He could not have done that! But we have a special problem, if we want to repeat what he might have done, for Herschel worked with a wide range of filters to hand – such as brandy in a decanter, for example. As one historian has noticed, his brandy was almost pitch black. We cannot repeat a measurement on that substance, whatever it was, today.

Herschel showed that heat and light are alike in reflection, refraction and differential refrangibility. He became troubled by transmission. He had the picture of a translucent medium stopping a definite proportion of the rays of a certain character, for example, red. His idea about red was that the heat ray, which refracts with the coefficient of red light, is identical to the red light with the same coefficient. So if $x\%$ of the light gets through, and heat and light are identical in this part of the spectrum, $x\%$ of the heat should go through too. He asks, ' Is the heat, which has the refrangibility of the red rays, occasioned by the light of those rays?' He finds not. A certain piece of glass that transmits nearly all the red light impedes 96.2% of the heat. Hence heat cannot be the same as light.

Herschel abandoned his original hypothesis and did not quite know what to think. Thus by the end of 1800, after 200 experiments and four major publications, he gave up. The very next year Thomas Young, whose work on interference commenced (or recreated) the wave theory of light, gave the Bakerian lecture in which he favoured Herschel's original hypothesis. Thus he was rather indifferent to Herschel's experimental dilemma. Perhaps the wave theory was more hospitable to radiant heat than was the Newtonian theory of rays of light particles. But in fact scepticism about radiant heat lasted long after Newtonian theory had gone into decline. It was resolved only by equipment invented by Macedonio Melloni (1798–1854). As soon as the thermocouple had been invented (1830) Melloni realized that he now had an instrument with which to measure the transmission of heat by different substances. This provides one of the innumerable examples in which an invention enables an experimenter to undertake another inquiry which in turn makes clear the route which the theoretician must follow.

Herschel had more primitive experimental problems. What was he observing? That was the question asked by his critics. He was rather viciously challenged in 1801. The experimental results were denied. A year later they were reproduced, more or less. There were many hard and simple experimental difficulties. For example, a prism does not neatly end at red. Some ambient light is diffused and comes below red as pale white light. So might not the 'infra-red' heat be caused by this white light? A new experimental idea intervened here. There is no significant invisible heat above purple, but might there not still be 'radiation'? It was known that silver chloride reacts when exposed at the purple end of the spectrum. (This is the beginning of photography.) Ritter exposed it beyond the violet and obtained a reaction; we now say that he discovered the ultraviolet in 1802.

On noticing

Herschel noticed the phenomenon of a differential heating by coloured light and reported this in as pure a sense-datum statement as we shall ever find in physics. I do not mean to discount the facts urged by N.R. Hanson, that one may see or notice a phenomenon only if one has a theory that makes sense of it. In Herschel's case it was lack of theory that made him sit up and take notice. Often we find the reverse. Hanson's book *The Positron* (1965), although containing some controversial accounts of discovery, is a sustained illustration of this thesis. He claims that people could see the tracks of positrons only when there was a theory, although after the theory, any undergraduate can see the selfsame tracks. We might call this the doctrine that noticing is theory-loaded.

Undoubtedly people tend to notice things that are interesting, surprising, and so forth, and such expectations and interests are influenced by theories they may hold – not that we should play down the possibility of the gifted 'pure' observer either. But there is a tendency to infer from stories like that of the positron, that anyone who reports, on looking at a photographic plate, 'that's a positron', is thereby implying or asserting a lot of theory. I do not think that this is so. An assistant can be trained to recognize those tracks without having a clue about the theory. In England it is still not too uncommon to find in a lab a youngish technician, with no formal education past 16 or 17, who is not only extraordinarily skilful with the apparatus, but also quickest at noting an oddity on for example the photographic plates he has prepared from the electron microscope.

But, it may be asked, is not the substance of the theory about positrons among the truth conditions or truth presuppositions for the type of utterance that we may represent by 'that's a positron'? Possibly, but I doubt it. The theory might be abandoned or superseded by a totally different theory about positrons, leaving intact what had, by then, become the class of observation sentences represented by 'that's a positron'. Of course the present theory might be wrecked in quite a different way, in which it turns out that so-called positron tracks are artifacts of the experimental device. That is only slightly more likely than the possibility that we shall discover that all sheep are only wolves in woolly suits. We would talk differently in that event too! I am not claiming that the sense of 'that's a positron' is any more unconnected to the rest of the discourse than 'that's a sheep'. I claim only that its sense need not be necessarily entangled in some particular theory, so that every time you say 'that's a positron' you somehow assert the theory.

Observation is a skill

An example similar to Hanson's makes the point that noticing and observation are skills. I think that Caroline Herschel (sister of William) discovered more comets than any other person in history. She got eight in a single year. Several things helped her do this. She was indefatigable. Every moment of cloudless night she was at her station. She also had a clever astronomer for a brother. She used a device, reconstructed only in 1980 by Michael Hoskin, that enabled her, each night, to scan the entire sky, slice by slice, never skimping on any corner of the heavens.[2] When she did find something curious 'with the naked eye', she had good telescopes to look more closely. But most important of all, she could recognize a comet at once. Everyone except possibly brother William had to follow the path of the suspected comet before reaching any opinion on its nature. (Comets have parabolic trajectories.)

In saying that Caroline Herschel could tell a comet just by looking, I do not mean to say that she was some mindless automaton. Quite the contrary. She had one of the deepest understandings of cosmology and one of the most profound speculative minds of her time. She was indefatigable not because she specially liked the boring task of sweeping the heavens, but because she wanted to know more about the universe.

It might well have turned out that Herschel's theory about comets was radically wrong. It might by now have been replaced by an account so different that some would call it incommensurable with hers. Yet this need not call in question her claim to fame. It would still be true that she discovered more comets than anyone else. Indeed if our new theory made comets into mere nothings, optical illusion on a cosmic scale, then her discovery of eight comets in a single year might bring more a smile of condescension than a gasp of admiration, but that is something else.

Seeing is not saying

The drive to displace observations by linguistic entities (observation sentences), persists throughout recent philosophy. Thus W.V.O. Quine proposes, almost as if it were a novelty, that we

2 M. Hoskin and B. Warner, 'Caroline Herschel's comet sweepers'. *Journal for the History of Astronomy* 12 (1981), pp. 27–34.

should 'drop the talk of observation and talk instead of observation sentences, the sentences that are said to report observations'. (*The Roots of Reference*, pp. 36–9.)

Caroline Herschel not only serves to rebut the claim that observation is just a matter of saying something, but also leads us to call in question the grounds for Quine's assertion. Quine was quite deliberately writing against the doctrine that all observations are theory-loaded. There is, he says, a perfectly distinguishable class of observation sentences, because 'observations are what witnesses will agree about, on the spot'. He assures us that a 'sentence is observational insofar as its truth value, on any occasion, would be agreed to by just about any member of the speech community witnessing the occasion'. And 'we can recognize membership in the speech community by mere fluency of dialogue'.

It is hard to imagine a more wrong-headed approach to observation in natural science. No one in Caroline Herschel's speech community would in general agree or disagree with her about a newly spotted comet, on the basis of one night's observation. Only she, and to a lesser extent William, had the requisite skill. This does not mean that we would say she had the skill unless other students, using other means, did not in the end come to agree on many of her identifications. Her judgements attain full validity only in the context of the rich scientific life of the period. But Quine's agreement 'on the spot' has little to do with observation in science.

If we want a comprehensive account of scientific life, we should, in exact opposition to Quine, drop the talk of observation sentences and speak instead of observation. We should talk carefully of reports, skills, and experimental results. We should consider what, for example, it is to have an experiment working well enough that the skilful experimenter knows that the data it provides may have some significance. What is it that makes an experiment convincing? Observation has precious little to do with that question.

Augmenting the senses

The unaided eye does not see very far or deep. Some of us need spectacles to avoid being practically blind. One way in which to extend the senses is by the use of ever more imaginative telescopes and microscopes. In the next chapter I discuss whether we see with a microscope (I think we do, but the issue is not simple). There are more radical extensions of the idea of observation. It is commonplace in the most rarefied reaches of experimental science to speak of 'observing' what we would naively suppose to be unobservable – if 'observable' really did mean, using the five senses almost unaided. Naturally if we were pre-positivist, like Bacon, we would say, 'so what?' But we still have a positivist legacy, and so we are a little startled by routine remarks by physicists. For example, the fermions are those fundamental particles with angular momentum such as $1/2$, or $3/2$, and which obey Fermi–Dirac statistics: they include electrons, nuons, neutrons, and protons, and much else, including the notorious quarks. One says things like: 'Of these fermions, only the t quark is yet unseen. The failure to observe tt' states in e^+e^- annihilation at PETRA remains a puzzle.[3]

The language which has been institutionalized among particle physicists may be seen by glancing at something as formal as a table of mesons. At the head of the April 1982 Meson Table one reads that 'quantities in italics are new or have been changed by more than one (old) standard deviation since April 1980'.[4] It is not clear even how to count the kinds of mesons which are now recorded, but let us limit ourselves to one open page (pp. 28–9) with nine mesons classified according to six different characteristics. Of interest is the 'partial decay mode' and the fraction of decays which are quantitatively recorded only when one has a statistical analysis at the 90% confidence level. Of the 31 decays associated with these nine mesons, we have 11 quantities or upper bounds, one entry 'large', one entry 'dominant', one entry '*dominant*', eight entries 'seen', six entries '*seen*', and three 'possibly seen'. Dudley Shapere has recently attempted a detailed analysis of such discourse.[5] He takes his example from talk of observing the interior of the sun, or another star, by collecting neutrinos in large quantities of cleaning fluid, and deducing various properties of the inside of the sun. Clearly this involves several layers, undreamt of by Bacon, of Bacon's idea of 'making manifest, things not directly perceptible, by means of others which are'. The trouble is that the physicist still calls this

3 C.Y. Prescott, 'Prospects for polarized electrons at high energies', Stanford Linear Accelerator, *SLAC-PUB-2630*, October 1980, p. 5. (This is a report connected with the experiment described in Chapter 16 below.)
4 *Particle Properties Data Booklet*, April 1982, p. 24. (Available from Lawrence Berkeley Laboratory and CERN. Cf. 'Review of physical properties', *Physics Letters* 111B (1982).)
5 D. Shapere, 'The concept of observation in science and philosophy', *Philosophy of Science* 49 (1982), pp. 231–67.

'direct observation'. Shapere has many quotations like these: 'There is no way known other than by neutrinos to see into a stellar interior.' 'Neutrinos,' writes another author, 'present the only way of directly observing' the hot stellar core.

Shapere concludes that this usage is apt and analyses it as follows: 'x is directly observed if (1) information is received by an appropriate receptor and (2) that information is transmitted directly, i.e. without interference, to the receptor from the entity x (which is the source of the information.)' I suspect that the usage of some physicists – illustrated by my quark quotation above – is even more liberal than this, but clearly Shapere gives the beginnings of a correct analysis.[6]

Massively theory-loaded observation (E)

Shapere notes that whether or not something is directly observable depends upon the current state of knowledge. Our theories of the workings of receptors, or of the transmission of information by neutrinos, all assume massive amounts of theory. So we might think that, as theory becomes taken for granted, we extend the realm of what we call observation. Yet we must never fall prey to the fallacy of talking about theory without making distinctions.

For example, there is an excellent reason for speaking of observation in connection with neutrinos and the sun. The theory of the neutrino and its interactions is almost completely independent of speculations about the core of the sun. It is precisely the disunity of science that allows us to observe (deploying one massive batch of theoretical assumptions) another aspect of nature (about which we have an unconnected bunch of ideas). Of course whether or not the two domains are connected itself involves, not exactly theory, but a hunch about the nature of nature. A slightly different example about the sun will illustrate this.

How might we investigate Dicke's hypothesis that the interior of the sun is rotating 10 times faster than its surface? Three methods have been proposed: (1) use optical observations of the oblateness of the sun; (2) try to measure the sun's quadruple mass-moment with the near fly-by of Starprobe, the satellite that goes within four solar radiuses of the sun; (3) measure the relativistic precession of a

---

6 See K.S. Shrader Frechette, 'Quark quantum numbers and the problem of microphysical observation', *Synthese* 50 (1982), pp. 125–46.

gyroscope in orbit about the sun. Do any of these three enable us to 'observe' interior rotation?

The first method assumes that optical shape is related to mass shape. A certain shape of the sun may help us infer something about internal rotation, but it is an inference based on an uncertain hypothesis which is itself connected with the subject matter under study.

The second method assumes that the only source of quadruple mass-moment is interior rotation, whereas it could be attributable to internal magnetic fields. Thus an assumption about what is going on (or not going on) in the sun itself is necessary for us to draw an inference about interior rotation.

On the other hand, relativistic precession of the gyroscope is based upon theory having nothing to do with the sun, and within the framework of present theory, one cannot conceive of anything except angular momentum of an object (e.g. the sun) that could produce such and such relativistic precession of a polar-orbiting gyro about the sun.

The point is not that the relativistic theory is better established than the theories involved in the other two possible experiments. Maybe relativistic precession theory will be the first to be abandoned. The point is that within the framework of our present understanding, the body of theoretical assumptions underlying the gyro proposal are arrived at in a completely different way from the propositions that people invent about the core of the sun. On the other hand, the first two proposals involve assumptions which in themselves concern beliefs about the sun's interior.

It is thus natural for the experimenter to say that the polar-orbiting gyro gives us a way to observe the interior rotation of the sun, while the other two investigations would only suggest inferences. This is not even to say that the third experiment would be the best one – its sheer cost and difficulty make the first two more attractive. I am making only a philosophical point about which experiments lead to observation, and which do not.

Possibly this connects with the debates about theory-loaded observation with which I began this chapter. Maybe the first two experiments contain theoretical assumptions connected with the subject under investigation, while the third, though loaded with theory, contains no such assumptions. In the case of seeing tables, our statements similarly contain no theoretical assumptions connected with the objects under inquiry, namely tables, even if (by an abuse of the words 'theory' and 'contain') they contain theoretical assumptions about vision.

Independence

On this view, something counts as observing rather than inferring when it satisfies Shapere's minimal criteria, and when the bundle of theories upon which it relies are not intertwined with the facts about the subject matter under investigation. The following chapter, on microscopes, confirms the force of this suggestion. I do not think that the issue is of much importance. Observation, in the philosophers' sense of producing and recording data, is only one aspect to experimental work. It is in another sense that the experimenter must be observant – sensitive and alert. Only the observant can make an experiment go, detecting the problems that are making it foul up, debugging it, noticing if something unusual is a clue to nature or an artifact of the machine. Such observation seldom appears in the finished reports of the experiment. It is at least as important as anything that does go into final write-ups, but nothing philosophical hangs on that.

Shapere had a more philosophical purpose in his analysis of observing. He holds that the old foundationalist view of knowledge was on the right track. Knowledge is in the end founded upon observation. He notes that what counts as observations depends upon our theories of the world and of special effects, so that there is no such thing as an absolute basic or observational sentence. But the fact that observing depends upon theories has none of the anti-rational consequences that have sometimes been inferred from the thesis that all observation is theory-loaded. Thus although Shapere has written the best extended study of observation in recent times, in the end he has an axe to grind, concerning the foundations for, and rationality of, theoretical belief. Van Fraassen also notes, in passing, that theory may delimit the bounds of observation. His purposes are different again. The real, for him, is observational, but he grants that theory itself can modify our beliefs about what is observational, and what is real. My purposes in this chapter have been more mundane. I have wanted to insist on some of the more humdrum aspects of observation. A philosophy of experimental science cannot allow theory-dominated philosophy to make the very concept of observation become suspect.

11 Microscopes

One fact about medium-size theoretical entities is so compelling an argument for medium-size scientific realism that philosophers blush to discuss it: Microscopes. First we guess there is such and such a gene, say, and then we develop instruments to let us see it. Should not even the positivist accept this evidence? Not so: the positivist says that only theory makes us suppose that what the lens teaches rings true. The reality in which we believe is only a photograph of what came out of the microscope, not any credible real tiny thing.

Such realism/anti-realism confrontations pale beside the meta-physics of serious research workers. One of my teachers, chiefly a technician trying to make better microscopes, could casually remark: 'X-ray diffraction microscopy is now the main interface between atomic structure and the human mind.' Philosophers of science who discuss realism and anti-realism have to know a little about the microscopes that inspire such eloquence. Even the light microscope is a marvel of marvels. It does not work in the way that most untutored people suppose. But why should a philosopher care how it works? Because it is one way to find out about the real world. The question is: How does it do it? The microscopist has far more amazing tricks than the most imaginative of armchair students of the philosophy of perception. We ought to have some understanding of those astounding physical systems 'by whose augmenting power we now see more/than all the world has ever done before'.[1]

The great chain of being

Philosophers have written dramatically about telescopes. Galileo himself invited philosophizing when he claimed to see the moons of Jupiter, assuming the laws of vision in the celestial sphere are the

---

[1] From a poem, 'In commendation of the microscope', by Henry Powers, 1664. Quoted in the excellent historical survey by Saville Bradbury, *The Microscope, Past and Present*, Oxford, 1968.

same as those on earth. Paul Feyerabend has used that very case to urge that great science proceeds as much by propaganda as by reason: Galileo was a con man, not an experimental reasoner. Pierre Duhem used the telescope to present his famous thesis that no theory need ever be rejected, for phenomena that don't fit can always be accommodated by changing auxiliary hypotheses (if the stars aren't where theory predicts, blame the telescope, not the heavens). By comparison the microscope has played a humble role, seldom used to generate philosophical paradox. Perhaps this is because everyone expected to find worlds within worlds here on earth. Shakespeare is merely an articulate poet of the great chain of being when he writes in *Romeo and Juliet* of Queen Mab and her minute coach 'drawn with a team of little atomies . . . her wag-goner, a small grey coated gnat not half so big as a round little worm prick'd from the lazy finger of a maid'. One expected tiny creatures beneath the scope of human vision. When dioptric glasses were to hand, the laws of direct vision and refraction went unquestioned. That was a mistake. I suppose no one understood how a microscope works before Ernst Abbe (1840–1905). One immediate reaction, by a president of the Royal Microscopical Society, and quoted for years in many editions of Gage's *The Microscope* – long the standard American textbook on microscopy – was that we do not, after all, see through a microscope. The theoretical limit of resolution

[A] Becomes explicable by the research of Abbe. It is demonstrated that microscopic vision is *sui generis*. There is and there can be *no* comparison between microscopic and macroscopic vision. The images of minute objects are not delineated microscopically by means of the ordinary laws of refraction; they are not dioptical results, but depend entirely on the laws of diffraction.

I think that this quotation, which I simply call [A] below, means that we do not see, in any ordinary sense of the word, with a microscope.

Philosophers of the microscope

Every twenty years or so a philosopher has said something about microscopes. As the spirit of logical positivism came to America, one could read Gustav Bergman telling us that as he used philosophical terminology, 'microscopic objects are not physical things in a literal sense, but merely by courtesy of language and pictorial imagination. . . . When I look through a microscope, all I see is a patch of color which creeps through the field like a shadow over a wall.'[2] In due course Grover Maxwell, denying that there is any fundamental distinction between observational and theoretical entities, urged a continuum of vision: 'looking through a window pane, looking through glasses, looking through binoculars, looking through a low power microscope, looking through a high power microscope, etc.'[3] Some entities may be invisible at one time and later, thanks to a new trick of technology, they become observable. The distinction between the observable and the merely theoretical is of no interest for ontology.

Grover Maxwell was urging a form of scientific realism. He rejected any anti-realism that holds that we are to believe in the existence of only the observable entities that are entailed by our theories. In *The Scientific Image* van Fraassen strongly disagrees. As we have seen in Part A above, he calls his philosophy constructive empiricism, and he holds that ' *Science aims to give us theories which are empirically adequate; and acceptance of a theory involves as belief only that it is empirically adequate*' (p. 12). Six pages later he attempts this gloss: 'To accept a theory is (for us) to believe that it is empirically adequate – that what the theory says *about what is observable* (by us) is true.' Clearly then it is essential for van Fraassen to restore the distinction between observable and unobservable. But it is not essential to him, exactly where we should draw it. He grants that 'observable' is a vague term whose extension itself may be determined by our theories. At the same time he wants the line to be drawn in the place which is, for him, most readily defensible, so that even if he should be pushed back a bit in the course of debate, he will still have lots left on the 'unobservable' side of the fence. He distrusts Grover Maxwell's continuum and tries to stop the slide from seen to inferred entities as early as possible. He quite rejects the idea of a continuum.

There are, says van Fraassen, two quite distinct kinds of case arising from Grover Maxwell's list. You can open the window and see the fir tree directly. You can walk up to at least some of the

2 G. Bergman, 'Outline of an empiricist philosophy of physics', *American Journal of Physics* 11 (1943), pp. 248–58, 335–42.
3 G. Maxwell, 'The ontological status of theoretical entities', in *Minnesota Studies in the Philosophy of Science* 3 (1962), pp. 3–27.

objects you see through binoculars, and see them in the round, with the naked eye. (Evidently he is not a bird watcher.) But there is no way to see a blood platelet with the naked eye. The passage from a magnifying glass to even a low powered microscope is the passage from what we might be able to observe with the eye unaided, to what we could not observe except with instruments. Van Fraassen concludes that we do not see through a microscope. Yet we see through some telescopes. We can go to Jupiter and look at the moons, but we cannot shrink to the size of a paramecium and look at it. He also compares the vapour trail made by a jet and the ionization track of an electron in a cloud chamber. Both result from similar physical processes, but you can point ahead of the trail and spot the jet, or at least wait for it to land, but you can never wait for the electron to land and be seen.

Don't just peer: interfere

Philosophers tend to regard microscopes as black boxes with a light source at one end and a hole to peer through at the other. There are, as Grover Maxwell puts it, low power and high power microscopes, more and more of the same kind of thing. That's not right, nor are microscopes just for looking through. In fact a philosopher will certainly not see through a microscope until he has learned to use several of them. Asked to draw what he sees he may, like James Thurber, draw his own reflected eyeball, or, like Gustav Bergman, see only 'a patch of color which creeps through the field like a shadow over a wall'. He will certainly not be able to tell a dust particle from a fruit fly's salivary gland until he has started to dissect a fruit fly under a microscope of modest magnification.

That is the first lesson: you learn to see through a microscope by doing, not just by looking. There is a parallel to Berkeley's *New Theory of Vision* of 1710, according to which we have three-dimensional vision only after learning what it is like to move around in the world and intervene in it. Tactile sense is correlated with our allegedly two-dimensional retinal image, and this learned cueing produces three-dimensional perception. Likewise a scuba diver learns to see in the new medium of the oceans only by swimming around. Whether or not Berkeley was right about primary vision, new ways of seeing, acquired after infancy, involve learning by doing, not just passive looking. The conviction that a particular part of a cell is there as imaged is, to say the least, reinforced when, using straightforward physical means, you microinject a fluid into just that part of the cell. We see the tiny glass needle – a tool that we have ourselves hand crafted under the microscope – jerk through the cell wall. We see the lipid oozing out of the end of the needle as we gently turn the micrometer screw on a large, thoroughly macroscopic, plunger. Blast! Inept as I am, I have just burst the cell wall, and must try again on another specimen. John Dewey's jeers at the 'spectator theory of knowledge' are equally germane for the spectator theory of microscopy.

This is not to say that practical microscopists are free from philosophical perplexity. Let us have a second quotation, [B], from the most thorough of available textbooks intended for biologists, E.M. Slayter's *Optical Methods in Biology*:

[B] The microscopist can observe a familiar object in a low power microscope and see a slightly enlarged image which is 'the same as' the object. Increase of magnification may reveal details in the object which are invisible to the naked eye; it is natural to assume that they, also, are 'the same as' the object. (At this stage it is necessary to establish that detail is not a consequence of damage to the specimen during preparation for microscopy.) But what is actually implied by the statement that 'the image is the same as the object?'

Obviously the image is a purely optical effect. . . . The 'sameness' of object and image in fact implies that the physical interactions with the light beam that render the object visible to the eye (or which would render it visible, if large enough) are identical with those that lead to the formation of an image in the microscope. . . .

Suppose however, that the radiation used to form the image is a beam of ultraviolet light, x-rays, or electrons, or that the microscope employs some device which converts differences in phase to changes in intensity. The image then cannot possibly be 'the same' as the object, even in the limited sense just defined! The eye is unable to perceive ultraviolet, x-ray, or electron radiation, or to detect shifts of phase between light beams. . . .

This line of thinking reveals that the image must be *a map of interactions between the specimen and the imaging radiation* (pp. 261–3).

The author goes on to say that all of the methods she has mentioned, and more, 'can produce "true" images which are, in some sense, "like" the specimen'. She also remarks that in a technique like the radioautogram 'one obtains an "image" of the specimen . . .

obtained exclusively from the point of view of the location of radioactive atoms. This type of "image" is so specialized as to be, generally, uninterpretable without the aid of an additional image, the photomicrograph, upon which it is superposed.'

This microscopist is happy to say that we see through a microscope only when the physical interactions of specimen and light beam are 'identical' for image formation in the microscope and in the eye. Contrast my quotation [A] from an earlier generation, and which holds that since the ordinary light microscope works by diffraction even it is not the same as ordinary vision but is *sui generis*. Can microscopists [A] and [B] who disagree about the simplest light microscope possibly be on the right philosophical track about 'seeing'? The scare quotes around 'image' and 'true' suggest more ambivalence in [B]. One should be especially wary of the word 'image' in microscopy. Sometimes it denotes something at which you can point, a shape cast on a screen, a micrograph, or whatever; but on other occasions it denotes as it were the input to the eye itself. The conflation results from geometrical optics, in which one diagrams the system with a specimen in focus and an 'image' in the other focal plane, where the 'image' indicates what you will see if you place your eye there. I do resist one inference that might be drawn even from quotation [B]. It may seem that any statement about what is seen with a microscope is theory-loaded: loaded with the theory of optics or other radiation. I disagree. One needs theory to make a microscope. You do not need theory to use one. Theory may help to understand why objects perceived with an interference-contrast microscope have asymmetric fringes around them, but you can learn to disregard that effect quite empirically. Hardly any biologists know enough optics to satisfy a physicist. Practice – and I mean in general doing, not looking – creates the ability to distinguish between visible artifacts of the preparation or the instrument, and the real structure that is seen with the microscope. This practical ability breeds conviction. The ability may require some understanding of biology, although one can find first class technicians who don't even know biology. At any rate physics is simply irrelevant to the biologist's sense of microscopic reality. The observations and manipulations seldom bear any load of physical theory at all, and what is there is entirely independent of the cells or crystals being studied.

Bad microscopes

I have encountered the impression that Leeuwenhoek invented the microscope, and that since then people have gone on to make better and better versions of the same kind of thing. I would like to correct that idea.

Leeuwenhoek, hardly the first microscopist, was a technician of genius. His microscopes had a single lens, and he made a lens for each specimen to be examined. The object was mounted upon a pin at just the right distance. We don't quite know how he made such marvellously accurate drawings of his specimens. The most representative collection of his lenses-plus-specimen was given to the Royal Society in London, which lost the entire set after a century or so in what are politely referred to as suspicious circumstances. But even by that time the glue for his specimens had lost its strength and the objects had begun to fall off their pins. Almost certainly Leeuwenhoek got his marvellous results thanks to a secret of illumination rather than lens manufacture, and he seems never to have taught the public his technique. Perhaps Leeuwenhoek invented dark field illumination, rather than the microscope. That guess should serve as the first of a long series of possible reminders that many of the chief advances in microscopy have had nothing to do with optics. We have needed microtomes to slice specimens thinner, aniline dyes for staining, pure light sources, and, at more modest levels, the screw micrometer for adjusting focus, fixatives and centrifuges.

Although the first microscopes did create a terrific popular stir by showing worlds within worlds, it is important to note that after Hooke's compound microscope, the technology did not markedly improve. Nor did much new knowledge follow after the excitement of the initial observations. The microscope became a toy for English ladies and gentlemen. The toy would consist of a microscope and a box of mounted specimens from the plant and animal kingdom. Note that a box of mounted slides might well cost more than the purchase of the microscope itself. You did not just put a drop of pond water on a slip of glass and look at it. All but the most expert would require a ready mounted slide to see *anything*. Indeed considering the optical aberrations it is amazing that anyone ever did see anything through a compound microscope, although in fact, as always in experimental science, a really skilful technician can do wonders with awful equipment.

There are about eight chief aberrations in bare-bones light microscopy. Two important ones are spherical and chromatic. The former is the result of the fact that you polish a lens by random rubbing. That, as can be proven, gives you a spherical surface. A light ray travelling at a small angle to the axis will not focus at the same point as a ray closer to the axis. For angles i for which $\sin i$ differs at all from i we get no common focus of the light rays, and so a point on the specimen can be seen only as a smear through the microscope. This was well understood by Huygens who also knew how to correct it in principle, but practical combinations of concave and convex lenses to avoid spherical aberration were a long time in the making.

Chromatic aberrations are caused by differences in wave length between light of different colours. Hence red and blue light emanating from the same point on the specimen will come to focus at different points. A sharp red image is superimposed on a blue smear or vice versa. Although rich people liked to have a microscope about the house for entertainments, it is no wonder that serious science had nothing to do with the instrument. We often regard Xavier Bichat as the founder of histology, the study of living tissues. In 1800 he would not allow a microscope in his lab. In the introduction to his *General Anatomy* he wrote that: 'When people observe in conditions of obscurity each sees in his own way and according as he is affected. It is, therefore, observation of the vital properties that must guide us', rather than the blurred images provided by the best of microscopes.

No one tried very hard to make achromatic microscopes, because Newton had written that they are physically impossible. They were made possible by the advent of flint glass, with refractive indices different from that of ordinary glass. A doublet of two lenses of different refractive indices can be made to cancel out the aberration perfectly for a given pair of red and blue wave lengths, and although the solution is imperfect over the whole spectrum, the result can be improved by a triplet of lenses. The first person to get the right ideas was so secretive that he sent the specifications for the lenses of different kinds of glass to two different contractors. They both subcontracted with the same artisan who then formed a shrewd guess that the lenses were for the same device. Hence, in 1758, the idea was pirated. A court case for the patent rights was decided in favour of the pirate, John Dolland. The High Court Judge ruled: 'It was not the person who locked the invention in his scritoire that ought to profit by a patent for such an invention, but he who brought it forth for the benefit of the public.'[4] The public did not benefit all that much. Even up into the 1860s there were serious debates as to whether globules seen through a microscope were artifacts of the instrument or genuine elements of living material. (They were artifacts.) Microscopes did get better and aids to microscopy improved at rather a greater rate. If we draw a graph of development we get a first high around 1660, then a slowly ascending plateau until a great leap around 1870; the next great period, which is still with us, commences about 1945. An historian has plotted this graph with great precision, using as a scale the limits of resolution of surviving instruments of different epochs. Making a subjective assessment of great applications of the microscope, we would draw a similar graph, except that the 1870/1660 contrast would be greater. Few truly memorable facts were found out with a microscope until after 1860. The surge of new microscopy is partly due to Abbe, but the most immediate cause of advance was the availability of aniline dyes for staining. Living matter is mostly transparent. The new aniline dyes made it possible for us to see microbes and much else.

Abbe and diffraction

How do we 'normally' see? Mostly we see reflected light. But if we are using a magnifying glass to look at a specimen illumined from behind, then it is transmission, or absorption, that we are 'seeing'. So we have the following idea: to see something through a light microscope is to see patches of dark and light corresponding to the proportions of light transmitted or absorbed. We see changes in the amplitude of light rays. I think that even Huygens knew there is something wrong with this conception, but not until 1873 did Abbe explain how a microscope works.

Ernst Abbe provides the happiest example of a rags to riches story. Son of a spinning-mill workman, he learned mathematics and

---

4 Quoted in Bradbury, *The Microscope, Past and Present*, p. 130.

was sponsored through the Gymnasium. He became a lecturer in mathematics, physics and astronomy. His optical work led him to be taken on by the small firm of Carl Zeiss in Jena, and when Zeiss died he became an owner; he retired to a life of philanthropy. Innumerable mathematical and practical innovations by Abbe turned Carl Zeiss into the greatest of optical firms. Here I consider only one.

Abbe was interested in resolution. Magnification is worthless if it 'magnifies' two distinct dots into one big blur. One needs to resolve the dots into two distinct images. It is a matter of diffraction. The most familiar example of diffraction is the fact that shadows of objects with sharp boundaries are fuzzy. This is a consequence of the wave character of light. When light travels between two narrow slits, some of the beam may go straight through, but some of it will bend off at an angle to the main beam, and some more will bend off at a larger angle: these are the first-order, second-order, etc., diffracted rays.

Abbe took as his problem how to resolve (i.e., visibly distinguish) parallel lines on a diatom (the tiny oceanic creatures that whales eat by the billion). These lines are very close together and of almost uniform separation and width. He was soon able to take advantage of even more regular artificial diffraction gratings. His analysis is an interesting example of the way in which pure science is applied, for he worked out the theory for the pure case of looking at a diatom or diffraction grating, and inferred that this represents the infinite complexity of the physics of seeing a heterogeneous object with a microscope.

When light hits a diffraction grating most of it is diffracted rather than transmitted. It is emitted from the grating at the angle of first-, second-, or third-order diffractions, where the angles of the diffracted rays are in part a function of the distances between the lines on the grating. Abbe realized that in order to see the slits on the grating, one must pick up not only the transmitted light, but also at least the first-order diffracted ray. What you see, in fact, is best represented as a Fourier synthesis of the transmitted and the diffracted rays. Thus according to Abbe the image of the object is produced by the interference of the light waves emitted by the principal image, and the secondary images of the light source which are the result of diffraction.

Practical applications abound. Evidently you will pick up more diffracted rays by having a wider aperture for the objective lens, but then you obtain vastly more spherical aberration as well. Instead you can change the medium between the specimen and the lens. With something denser than air, as in the oil-immersion microscope, you capture more of the diffracted rays within a given aperture and so increase the resolution of the microscope.

Although the first Abbe–Zeiss microscopes were good, the theory was resisted for a number of years, particularly in England and America, who had enjoyed a century of dominating the market. Even by 1910 the very best English microscopes, built on purely empirical experience, although stealing a few ideas from Abbe, could resolve as well or better than the Zeiss equipment. This is not entirely unusual. Although sailing ships have been part of human culture almost for ever, the greatest improvements in the sailing ship were made between 1870 and 1900, when the steamboat had made them obsolete. It was just at that time that craftsmanship peaked. Likewise with the microscope, but of course the expensive untheoretical English craftsmen of microscopy were as doomed as the sailing ship.

It was not, however, only commercial or national rivalry which made some people hesitate to believe in Abbe. I noted above that quotation [A] is used in Gage's *The Microscope.* In the ninth edition (1901) of that textbook the author refers to the alternative theory that microscopic vision is the same 'with the unaided eye, the telescope and the photographic camera. This is the original view, and the one which many are favoring at the present day.' In the 11th edition (1916) this is modified: 'Certain very striking experiments have been devised to show the accuracy of Abbe's hypothesis, but as pointed out by many, the ordinary use of the microscope never involves the conditions realized in these experiments.' This is a fine example of what Lakatos calls a degenerating research programme. The passage remains the same, in essentials, even in the 17th edition (1941). Thus there was a truly deep-seated repugnance to Abbe's doctrine which, as quotation [A] has it, says 'there is and can be *no* comparison between microscopic and macroscopic vision'.

If you hold (as my more modern quotation [B] still seems to hold), that what we see is essentially a matter of a certain sort of physical processing in the eye, then everything else must be more in the domain of optical illusion or at best of mapping. On that account, the systems of Leeuwenhoek and of Hooke do allow you to see. After Abbe even the conventional light microscope is essentially a Fourier synthesizer of first- or even second-order diffractions. Hence you must modify your notion of seeing or hold that you never see through a serious microscope. Before reaching a conclusion on this question, we had best examine some more recent instruments.

A plethora of microscopes

We move on to after World War II. Most of the ideas had been around during the interwar years, but did not get beyond proto-types until later. One invention is a good deal older, but it was not properly exploited for a while.

The first practical problem for the cell biologist is that most living material does not show up under an ordinary light microscope because it is transparent. To see anything you have to stain the specimen. Most aniline dyes are number one poisons, so what you will see is usually a very dead cell, which is also quite likely to be a structurally damaged cell, exhibiting structures that are an artifact of the preparation. However it turns out that living material varies in its birefringent (polarizing) properties. So let us incorporate into our microscope a polarizer and an analyser. The polarizer transmits to the specimen only polarized light of certain properties. In the simplest case, let the analyser be placed at right angles to the polarizer, so as to transmit only light of polarization opposite to that of the polarizer. The result is total darkness. But suppose the specimen is itself birefringent; it may then change the plane of polarization of the incident light, and so a visible image may be formed by the analyser. Transparent fibers of striated muscle may be observed in this way, without any staining, and relying solely on certain properties of light that we do not normally 'see'.

Abbe's theory of diffraction, augmented by the polarizing microscope, leads to something of a conceptual revolution. We do not need the 'normal' physics of seeing in order to perceive structures in living material. In fact we seldom use it. Even in the standard case we synthesize diffracted rays rather than seeing the specimen by way of 'normal' visual physics. The polarizing microscope reminds us that there is more to light than refraction, absorption and diffraction. We could use any property of light that interacts with a specimen in order to study the structure of the specimen. Indeed we could use any property of *any kind of wave* at all.

Even when we stick to light there is lots to do. Ultraviolet microscopy doubles resolving power, although its chief interest lies in noting the specific ultraviolet absorptions that are typical of certain biologically important substances. In fluorescence microscopy the incident illumination is cancelled out, and one observes only light re-emitted at different wave lengths by natural or induced phosphorescence or fluorescence. This is an invaluable histological technique for certain kinds of living matter. More interesting, however, than using unusual modes of light transmission or emission, are the games we can play with light itself: the Zernicke phase contrast microscope and the Nomarski interference microscope.

A specimen that is transparent is uniform with respect to light absorption. It may still possess invisible differences in refractive index in various parts of its structure. The phase contrast microscope converts these into visible differences of intensity in the image of the specimen. In an ordinary microscope the image is synthesized from the diffracted waves D and the directly transmitted waves U. In the phase contrast microscope the U and D waves are physically separated in an ingenious although physically simple way, and one or the other kind of wave is then subject to a standard phase delay which has the effect of producing in focus phase contrasts corresponding to the differences in refractive index in the specimen.

The interference contrast microscope is perhaps easier to understand. The light source is simply split by a half silvered mirror, and half the light goes through the specimen while half is kept as an unaffected reference wave to be recombined for the output image. Changes in optical path due to different refractive indices within the specimen thus produce interference effects with the reference beam.

The interference microscope is attended by illusory fringes but is particularly valuable because it provides a quantitative determination of refractive indices within the specimen. Naturally once we have such devices in hand, endless variations may be constructed, such as polarizing interference microscopes, multiple beam interference, phase modulated interference and so forth.

Theory and grounds for belief

Some theory of light is of course essential for building a new kind of microscope, and is usually important for improving an old kind. Interference or phase contrast microscopes could hardly have been invented without a wave theory of light. The theory of diffraction helped Abbe and his company make better microscopes. We should not, however, underestimate the pre-theoretical role of invention and fiddling around. For a couple of decades the old empirical microscope manufacturers made better microscopes than Zeiss. When the idea of an electron microscope was put into practice, it was a long shot, because people were convinced, on theoretical grounds, that the specimen would almost instantly be fried and then burnt out. The X-ray microscope has been a theoretical possibility for ages, but can effectively be built only in the next few years using high quality beams that can be bought from a linear accelerator. Likewise the acoustic microscope described below has long been an obvious possibility, but only in the last 10 years has one had the fast electronics to produce good high frequency sound and quality scanners. Theory has had only a modest amount to do with building these ingenious devices. The theory involved is mostly of the sort you learn in Physics I at college. It is the engineering that counts.

Theory may seem to enter at another level. Why do we believe the pictures we construct using a microscope? Is it not because we have a theory according to which we are producing a truthful picture? Is this not yet another case of Shapere's remark, that what we call observation is itself determined by theory? Only partially. Despite Bichat, people rightly believed much of what they saw through pre-Abbe microscopes, although they had only the most inadequate and commonplace theory to back them up (wrongly, as it happened). Visual displays are curiously robust under changes of theory. You produce a display, and have a theory about why a tiny specimen looks like that. Later you reverse the theory of your microscope, and you still believe the representation. Can theory really be the source of our confidence that what we are seeing is the way things are?

In correspondence Heinz Post told me that long ago he had discussed the field emission microscope in order to illustrate the importance of producing visual representations of large molecules. (His example concerned anthracene rings.) At the time, this device was taken to confirm what F.A. Kekule (1829–96) had postulated in

1865, that the benzene molecules are rings involving six carbon atoms. The original theory about the field emission microscope was that one was seeing essentially shadows of the molecules, that is, that we were observing an absorption phenomenon. Post learned much later that the underlying theory had been reversed. One was observing diffraction phenomena. It made no whit of difference. People kept on regarding the micrographs of the molecules as genuinely correct representations. Is this all mumbo-jumbo, a sort of confidence trick? Only a theory-dominated philosophy would make one think so. The experimental life of microscopy uses non-theory to sort out artifacts from the real thing. Let us see how it goes.

Truth in microscopy

The differential interference-contrast technique is distinguished by the following characteristics: Both clearly visible outlines (edges) within the object and continuous structures (striations) are imaged in their true profile.

So says a Carl Zeiss sales catalogue to hand. What makes the enthusiastic sales person suppose that the images produced by these several optical systems are 'true'? Of course, the images are true only when one has learned to put aside distortions. There are many grounds for the conviction that a perceived bit of structure is real or true. One of the most natural is the most important. I shall illustrate it with my own first experience in the laboratory. Low powered electron microscopy reveals small dots in red blood platelets. These are called dense bodies: that means simply that they are electron dense, and show up on a transmission electron microscope without any preparation or staining whatsoever. On the basis of the movements and densities of these bodies in various stages of cell development or disease, it is guessed that they may have an important part to play in blood biology. On the other hand they may simply be artifacts of the electron microscope. One test is obvious: can one see these selfsame bodies using quite different physical techniques? In this case the problem is fairly readily solved. The low resolution electron microscope is about the same power as a high resolution light microscope. The dense bodies do not show up under every technique, but are revealed by fluorescent staining and subsequent observation by the fluorescent microscope.

Slices of red blood platelets are fixed upon a microscopic grid.

This is literally a grid: when seen through the microscope one sees a grid each of whose squares is labelled with a capital letter. Electron micrographs are made of the slices mounted upon such grids. Specimens with particularly striking configurations of dense bodies are then prepared for fluorescence microscopy. Finally one compares the electron micrographs and the fluorescence micrographs. One knows that the micrographs show the same bit of the cell, because this bit is clearly in the square of the grid labelled *P*, say. In the fluorescence micrographs there is exactly the same arrangement of grid, general cell structure and of the 'bodies' seen in the electron micrograph. It is inferred that the bodies are not an artifact of the electron microscope.

Two physical processes – electron transmission and fluorescent re-emission – are used to detect the bodies. These processes have virtually nothing in common between them. They are essentially unrelated chunks of physics. It would be a preposterous coincidence if, time and again, two completely different physical processes produced identical visual configurations which were, however, artifacts of the physical processes rather than real structures in the cell.

Note that no one actually produces this 'argument from coincidence' in real life. One simply looks at the two (or preferably more) sets of micrographs from different physical systems, and sees that the dense bodies occur in exactly the same place in each pair of micrographs. That settles the matter in a moment. My mentor, Richard Skaer, had in fact expected to prove that dense bodies are artifacts. Five minutes after examining his completed experimental micrographs he knew he was wrong.

Note also that no one need have any ideas what the dense bodies *are*. All we know is that there are some structural features of the cell rendered visible by several techniques. Microscopy itself will never tell all about these bodies (if indeed there is anything important to tell). Biochemistry must be called in. Also, instant spectroscopic analysis of the dense body into constituent elements is now available, by combining an electron microscope and a spectroscopic analyser. This works much like spectroscopic analyses of the stars.

Coincidence and explanation

This argument from coincidence may seem like a special case of the cosmic accident argument mentioned at the end of Chapter 3.

Theories explain diverse phenomena, and it would be a cosmic accident if a theory were false and yet correctly predicted the phenomena. We 'infer to the best explanation' that the theory is true. The common cause of the phenomena must be the theoretical entities postulated by the theory. As an argument for scientific realism this idea has produced much debate. So it may seem as if my talk of coincidence puts me in the midst of an ongoing feud. Not so! My argument is much more localized.

First of all such arguments are often put in terms of an observational vocabulary and a theoretical one. ('Innumerable lucky accidents bringing about the behaviour mentioned in the observational vocabulary, *as if* they were brought about by the nonexistent things talked about in the theoretical vocabulary.') Well, we are not concerned with an observational and theoretical vocabulary. There may well be no theoretical vocabulary for the things seen under the microscope – 'dense body' means nothing else than something dense, that is, something that shows up under the electron microscope without any staining or other preparation. Secondly we are not concerned with explanation. We see the same constellations of dots whether we use an electron microscope or fluorescent staining, and it is no 'explanation' of this to say that some definite kind of thing (whose nature is as yet unknown) is responsible for the persistent arrangements of dots. Thirdly we have no theory which predicts some wide range of phenomena. The fourth and perhaps most important difference is this: we are concerned to distinguish artifacts from real objects. In the metaphysical disputes about realism, the contrast is between 'real although unobservable entity' and 'not a real entity, but rather a tool of thought'. With the microscope we know there are dots on the micrograph. The question is, are they artifacts of the physical system or are they structure present in the specimen itself? My argument from coincidence says simply that it would be a preposterous coincidence if two totally different kinds of physical systems were to produce exactly the same arrangements of dots on micrographs.

The argument of the grid

I now venture a philosopher's aside on the topic of scientific realism. Van Fraassen says we can see through a telescope because although we need the telescope to see the moons of Jupiter when we are positioned on earth, we could go out there and look at the moons with the naked eye. That is not so fanciful as it sounds, for there is a very small number of people living today who, it appears, can distinguish Jupiter's moons with the naked eye from here. For those of us with less acuity it is, for the moment however, science fiction. The microscopist avoids fantasy. Instead of flying to Jupiter we shrink the visible world. Consider the grids used to re-identify dense bodies. The tiny grids are made of metal; they are barely visible to the naked eye. They are made by drawing a very large grid with pen and ink. Letters are neatly inscribed at the corner of each square on the grid. Then the grid is reduced photographically. Using what are now standard techniques, metal is deposited on the resulting micrograph. Grids are sold in packets, or rather tubes, of 100, 250 and 1000. The procedures for making such grids are entirely well understood, and as reliable as any other high quality mass production system.

In short, rather than disporting ourselves to Jupiter in an imaginary space ship, we are routinely shrinking a grid. Then we look at the tiny disc through almost any kind of microscope and see exactly the same shapes and letters as were originally drawn on a large scale. It is impossible seriously to entertain the thought that the minute disc, which I am holding by a pair of tweezers, does not in fact have the structure of a labelled grid. I know that what I see through the microscope is veridical because we *made* the grid to be just that way. I know that the process of manufacture is reliable, because we can check the results with the microscope. Moreover we can check the results with any kind of microscope, using any of a dozen unrelated physical processes to produce an image. Can we entertain the possibility that, all the same, this is some gigantic coincidence? Is it false that the disc is, microscopically, in the shape of a labelled grid? Is it a gigantic conspiracy of 13 totally unrelated physical processes that the large scale grid was shrunk into some non-grid which when viewed using 12 different kinds of microscopes still looks like a grid? To be an anti-realist about that grid you would have to invoke a malign Cartesian demon of the microscope.

The argument of the grid requires a healthy recognition of the disunity of science, at least at the phenomenological level. Light microscopes, trivially, all use light, but interference, polarizing, phase contrast, direct transmission, fluorescence and so forth exploit essentially unrelated phenomenological aspects of light. If the same structure can be discerned using many of these different aspects of light waves, we cannot seriously suppose that the structure is an artifact of all the different physical systems. Moreover I emphasize that all these physical systems are made by people. We purify some aspect of nature, isolating, say, the phase interference character of light. We design an instrument knowing in principle exactly how it will work, just because optics is so well understood a science. We spend a number of years debugging several prototypes, and finally have an off-the-shelf instrument, through which we discern a particular structure. Several other off-the-shelf instruments, built upon entirely different principles, reveal the same structure. No one short of the Cartesian sceptic can suppose that the structure is made by the instruments rather than inherent in the specimen.

In 1800 it was not only possible but perfectly sensible to ban the microscope from the histology lab on the plain grounds that it chiefly revealed artifacts of the optical system rather than the structure of fibres. That is no longer the case. It is always a problem in innovative microscopy to become convinced that what you are seeing is really in the specimen rather than an artifact of the preparation of the optics. But in 1983, as opposed to 1800, we have a vast arsenal of ways of gaining such conviction. I emphasize only the 'visual' side. Even there I am simplistic. I say that if you can see the same fundamental features of structure using several different physical systems, you have excellent reason for saying, 'that's real' rather than, 'that's an artifact'. It is not conclusive reason. But the situation is no different from ordinary vision. If black patches on the tarmac road are seen, on a hot day, from a number of different perspectives, but always in the same location, one concludes that one is seeing puddles rather than the familiar illusion. One may still be wrong. One is wrong, from time to time, in microscopy too. Indeed the sheer similarity of the kinds of mistakes made in macroscopic and microscopic perception may increase the inclination to say, simply, that one sees through a microscope.

I must repeat that just as in large scale vision, the actual images or micrographs are only one small part of the confidence in reality. In a recent lecture the molecular biologist G.S. Stent recalled that in the late forties *Life* magazine had a full colour cover of an electron micrograph, labelled, excitedly, 'the first photograph of the gene' (March 17 1947). Given the theory, or lack of theory, of the gene at that time, said Stent, the title did not make any sense. Only a greater understanding of what a gene is can bring the conviction of what the micrograph shows. We become convinced of the reality of bands and interbands on chromosomes not just because we see them, but because we formulate conceptions of what they do, what they are for. But in this respect too, microscopic and macroscopic vision are not different: a Laplander in the Congo won't see much in the bizarre new environment until he starts to get some idea what is in the jungle.

Thus I do not advance the argument from coincidence as the sole basis of our conviction that we see true through the microscope. It is one element, a compelling visual element, that combines with more intellectual modes of understanding, and with other kinds of experimental work. Biological microscopy without practical bio-chemistry is as blind as Kant's intuitions in the absence of concepts.

The acoustic microscope

I here avoid the electron microscope. There is no more 'the' electron microscope than 'the' light microscope: all sorts of different properties of electron beams are used. This is not the place to explain all that, but in case we have in mind too slender a diet of examples based upon the properties of visible light, let us briefly consider the most disparate kind of radiation imaginable: sound.[5]

Radar, invented for aerial warfare, and sonar, invented for war at sea, remind us that longitudinal and transverse wave fronts can be put to the same kinds of purpose. Ultrasound is 'sound' of very high frequency. Ultrasound examination of the foetus in the mother's womb has recently won well deserved publicity. Over 40 years ago Soviet scientists suggested a microscope using sound of frequency 1000 times greater than audible noise. Technology has only recently caught up to this idea. Useful prototypes are just now in operation.

The acoustic part of the microscope is relatively simple. Electric signals are converted into sound signals and then, after interaction with the specimen, are reconverted into electricity. The subtlety of

5 See, for example, C. F. Quate, 'The acoustic microscope', *Scientific American* 241 (Oct. 1979), pp. 62–9.

present instruments lies in the electronics rather than the acoustics. The acoustic microscope is a scanning device. It produces its images by converting the signals into a spatial display on a television screen, a micrograph, or, when studying a large number of cells, a videotape.

As always a new kind of microscope is interesting because of the new aspects of a specimen that it may reveal. Changes in refractive index are vastly greater for sound than for light. Moreover sound is transmitted through objects that are completely opaque. Thus one of the first applications of the acoustic microscope is in metallurgy, and also in detecting defects in silicon chips. For the biologist, the prospects are also striking. The acoustic microscope is sensitive to density, viscosity and flexibility of living matter. Moreover the very short bursts of sound used by the scanner do not immediately damage the cell. Hence one may study the life of a cell in a quite literal way: one will be able to observe changes in viscosity and flexibility as the cell goes about its business.

The rapid development of acoustic microscopy leaves us uncertain where it will lead. A couple of years ago the research reports carefully denied any competition with electron microscopes; they were glad to give resolution at about the level of light microscopes. Now, using the properties of sound in supercooled solids one can emulate the resolution of electron microscopes, although that is not much help to the student of living tissue!

Do we see with an acoustic microscope?

Looking with a microscope

Looking through a lens was the first step in technology. Then came peering through the tube of a compound microscope, but looking 'through' the instrument is immaterial. We study photographs taken with a microscope. Thanks to the enormous depth of focus of an electron microscope it is natural to view the image on a large flat surface so everyone can stand around and point to what's interesting. Scanning microscopes necessarily constitute the image on a screen or plate. Any image can be digitized and retransmitted on a television display or whatever. Moreover, digitization is marvellous for censoring noise and even reconstituting lost information. Do not, however, become awed by technology. In the study of crystal structure, one good way to get rid of noise is to cut up a micrograph in a systematic way, paste it back together, and rephotograph it for interference contrast. Thus we do not in general see through a microscope; we see with one. But do we *see* with a microscope? It would be silly to debate the ordinary use of the word 'see', especially given the usages quoted at the end of the last chapter, where we 'see' most of the fermions, or 'observe' the sun's core with neutrinos. Consider a device for low-flying jet planes, laden with nuclear weapons, skimming a few dozen yards from the surface of the earth in order to evade radar detection. The vertical and horizontal scale are both of interest to the pilot who needs both to see a few hundred feet down and miles and miles away. The visual information is digitized, processed, and cast on a head-up display on the windscreen. The distances are condensed and the altitude is expanded. Does the pilot see the terrain? Yes. Note that this case is not one in which the pilot could have seen the terrain by getting off the plane and taking a good look. There is no way to look at that much landscape without an instrument.

Consider the electron diffraction microscope with which I produce images of crystals in either conventional or reciprocal space – nowadays, at the flick of a switch. Because the dots of an electron diffraction pattern are reciprocal to the atomic structure of a crystal, reciprocal space is, roughly speaking, conventional space turned inside out. Near is far and far is near. Crystallographers often find it most natural to study their specimens in reciprocal space. Do they see them in reciprocal space? They certainly say so, and thereby call in question the Kantian doctrine of the uniqueness of perceptual space.

How far could one push the concept of seeing? Suppose I take an electronic paint brush and paint, on a television screen, an accurate picture (a) of a cell that I have previously studied, say, by using a digitized and reconstituted image (b). Even if I am 'looking at the cell' in case (b), in (a) I am only looking at a drawing of the cell. What is the difference? The important feature is that in (b) there is a direct interaction between a wave source, an object, and a series of physical events that end up in an image of the object. To use quotation [B] once again, in case (b) we have a map of interactions between the specimen and the imaging radiation. If the map is a good one, then (b) is seeing with a microscope.

This is doubtless a liberal extension of the notion of seeing. We see with an acoustic microscope. We see with television, of course. We do not say that we saw an attempted assassination *with* the television, but *on* the television. That is mere idiom, inherited from 'I heard it on the radio.' We distinguish between seeing the television broadcast live or not. We have endless distinctions to be made with various adverbs, adjectives and even prepositions. I know of no confusion that will result from talk of seeing with a microscope.

Scientific realism

When an image is a map of interactions between the specimen and the image of radiation, and the map is a good one, then we are seeing with a microscope. What is a good map? After discarding or disregarding aberrations or artifacts, the map should represent some structure in the specimen in essentially the same two- or three-dimensional set of relationships as are actually present in the specimen.

Does this bear on scientific realism? First let us be clear that it can bear in only the modest way. Imagine a reader initially attracted by van Fraassen, and who thought that objects seen only with light microscopes do not count as observable. That reader could change his mind, and admit such objects into the class of observable entities. This would still leave intact all the main philosophical positions of van Fraassen's anti-realism.

But if we conclude that we see with the light microscopes, does it follow that the objects we report seeing are real? No. For I have said only that we should not be stuck in the nineteenth-century rut of positivism-cum-phenomenology, and that we should allow ourselves to talk of seeing with a microscope. Such a recommendation implies a strong commitment to realism about microscopy, but it begs the question at issue. This is clear from my quotation from high-energy physics, with its cheerful talk of our having seen electron neutrinos and so forth. The physicist is a realist too, and he shows this by using the word 'see', but his usage is no *argument* that there are such things.

Does microscopy then beg the question of realism? No. We *are* convinced of the structures that we observe using various kinds of microscopes. Our conviction arises partly from our success at systematically removing aberrations and artifacts. In 1800 there was no such success. Bichat banned the microscope from his dissecting rooms, for one did not, then, observe structures that could be confirmed to exist in the specimens. But now we have by and large got rid of aberrations; we have removed many artifacts, disregard others, and are always on the lookout for undetected frauds. We are convinced about the structures we seem to see because we can interfere with them in quite physical ways, say by microinjecting. We are convinced because instruments using entirely different physical principles lead us to observe pretty much the same structures in the same specimen. We are convinced by our clear understanding of most of the physics used to build the instruments that enable us to see, but this theoretical conviction plays a relatively small part. We are more convinced by the admirable intersections with biochemistry, which confirm that the structures that we discern with the microscope are individuated by distinct chemical properties too. We are convinced not by a high powered deductive theory about the cell – there is none – but because of a large number of interlocking low level generalizations that enable us to control and create phenomena in the microscope. In short, we learn to move around in the microscopic world. Berkeley's *New Theory of Vision* may not be the whole truth about infantile binocular three-dimensional vision, but is surely on the right lines when we enter the new worlds within worlds that the microscope reveals to us.

12 Speculation, calculation, models, approximations

I have now discouraged the idea that there is just one monolithic practice, observing. We must now apply the same tactics to the other side of that old duet of theory and observation. Theory is no more one kind of thing than observation is. A rich but elementary example will illustrate this fact.

The Faraday effect

Michael Faraday (1791–1867), an apprentice bookbinder, got a job when he was 21 as assistant to Humphry Davy. He then advanced our knowledge and transformed our machinery. His two most lasting insights go hand in hand: the invention of the electric motor (and, conversely, the electric dynamo); and the realization that changes in current produce changes in magnetic intensity (conversely, rotation through a magnetic field generates current). There is also what is called the Faraday effect, or the magneto-optical effect. Faraday found that magnetism can affect light. This is of enormous historical importance. It suggested that there might be a single theory unifying light and electromagnetism. James Clerk Maxwell put it together by 1861, and systematically presented it in 1873. Faraday's effect had been experimentally demonstrated in 1845.

Faraday, a deeply religious man, was convinced that all the forces of nature must be interconnected. Newton made a space for unified science that lasted until 1800. In that year, as we saw in Chapter 10, William Herschel produced the problem of radiant heat. In the same year Guiseppe Volta made the first voltaic cell. There was, for the first time, a source of a steady electric current, which, as Øersted soon showed, could affect the needle of a magnetic compass. In 1801 Thomas Young announced the wave theory of light, putting paid to a century of Newtonian ray-theory of light. In short, the Newtonian unity of science was in shambles. Moreover, there was no apparent connection between the forces of electromagnetism, of gravity, of light. Michael Faraday addressed himself to this question. David Brewster, the great experimentalist mentioned in Chapter 9, had shown in 1819 that by putting a strain on some kinds of glass you could make the glass polarize light. Using this analogy, Faraday guessed that if stressing a body could affect the transmission of light, electrifying it might do so too. Faraday tried to find such an effect repeatedly, in 1822, in 1834, in 1844. Then in 1845 he gave up electrification and tried magnetism instead. Even this was a failure until he used dense glass he had developed many years earlier for another purpose. He found that the plane of polarization of a beam of light would rotate when sent through this borosylicate glass, parallel to the lines of magnetic force. The French physicist M.E. Verdet (1824–96) later explored this property in a wide range of substances, thereby establishing it as a general characteristic of nature.

Explaining Faraday's effect (E)

Faraday had no theory of what he had found. In the next year, 1846, G.B. Airy (1801–92) showed how to represent it analytically within the wave theory of light. The equations for light had contained some second derivatives of displacement with respect to time. Airy added some *ad hoc* further terms, either first or third derivatives. This is a standard move in physics. In order to make the equations fit the phenomena, you pull from the shelf some fairly standard extra terms for the equations, without knowing why one rather than another will do the trick.

In 1856 Kelvin proposed a physical model: the magnetic field makes the molecules in the glass block rotate about axes parallel to the lines of force. These molecular rotations couple to the vibrations induced by the light waves, and thereby make the plane of polarization rotate.

Kelvin's model was adapted by Maxwell and helped form his electromagnetic theory of light. However it did not agree well with experimental details reported by Verdet. So Maxwell used symmetry arguments to determine the additional terms in the Lagrangian of the electromagnetic field vector which is used to describe the phenomena. Finally in 1892 H.A. Lorentz combined Maxwell's equations with his electron theory. This gave the explanation used today. The effect is accounted for physically – Kelvin style – by a local motion around the lines of force. But it is not a Kelvinian mysterious molecular rotation that just happens. It is a motion of electrons induced electromagnetically.

Six levels of 'theory'

Our story illustrates at least six different levels of theory. They are not merely levels of greater generality or deductive power, but rather different kinds of speculation. The basic experimental work is that of Faraday followed by Verdet. The 'theoretical' ideas, in order of appearance, are as follows:

1 Motivated by faith in the unity of science, Faraday speculates that there *must* be some connection between electromagnetism and light.

2 There is Faraday's analogy with Brewster's discovery: something electromagnetic may affect polarizing properties.

3 Airy provides an *ad hoc* mathematical representation.

4 Kelvin gives a physical model, using a mechanical picture of rotating molecules in glass.

5 Maxwell uses symmetry arguments to provide a formal analysis within the new electromagnetic theory.

6 Lorentz provides a physical explanation within electron theory.

I do not mean to imply that these different kinds of hypotheses occur in connection with all research, nor that they need occur in this order. This rather Baconian history begins with a broad idea and an analogy; it is substantiated by experiment, and then is developed into increasingly more satisfactory theoretical formulations. Often, of course, the big speculation (6) comes first. The example illustrates only the humdrum but easy-to-forget fact that 'theory' covers lots of productions. A dictionary says that etymologically, the word 'theory' is derived from a Greek word which, in one connotation, is a speculation. Let us fix on that.

Speculation

Rather than the simple dichotomy, C.W.F. Everitt and I prefer a tripartite division of activities. I call it speculation, calculation and experimentation.

The word 'speculation' can apply to all sorts of waffling and stock-marketeering. By speculation I shall mean the intellectual representation of something of interest, a playing with and re-structuring of ideas to give at least a qualitative understanding of some general feature of the world.

Are speculations only qualitative? Of course not. Physics is a quantitative science. Yet most theories have some free parameters that are filled in by experiment. The underlying theory is more qualitative. One old speculation is that the distance traversed by a body falling freely towards the earth varies as the square of the time taken to fall. This is represented as $1/2\ gt^2$. The numerical value of the local acceleration of gravity, g, is no part of the initial speculation. It is just a blank, that we fill in by non-theoretical measurement. At present all quantitative theory says in the end: 'The equations are of the form so and so, with certain constants of nature to be filled in, empirically.' There has long been a Leibnizian dream of explaining away fundamental constants, but that is still an exciting programme, not a field with results. Thus for all its paraphernalia of quantity, speculation may be essentially qualitative.

There are at least as many kinds of speculation as there are kinds of representation. There are physical models, illustrated by Kelvin's account of the Faraday effect. There are mathematical structures. Both approaches have led to remarkable insights. According to one misleading cliché about late-nineteenth-century science, German physicists used primarily mathematical approaches while British ones made physical models. Both kinds of work collaborate, and both kinds of worker often uncovered almost the same facts in quite different ways. Moreover, on closer inspection most of the physical modelling, of for example Maxwell, turns out to involve abstract structures. Thus the elements of his statistical mechanics were not hard particles but mathematical differentials with no evident physical meaning. Conversely much of the applied mathematics in Germany hinged on description of plain physical models. These aspects of the human mind are not in general separable, but will continue to be permuted and altered in ways which we cannot foresee.

Calculation

Kuhn remarks that normal science is a matter of what he calls *articulation*. We articulate theory to make it mesh better with the world, open to experimental verification. Most initial speculations hardly mesh with the world at all. This is for two reasons. One is that one can seldom directly deduce from a speculation consequences that are even in principle testable. The other is that even a proposition which is in principle testable is often not testable, simply because no one knows how to conduct the test. New experimental ideas and new kinds of technology are required. In the example of Herschel and radiant heat, we required the thermocouple and ideas of Macedonio Melloni in order to dig into Herschel's initial speculations.

Thus Kuhn's articulation must denote two kinds of thing, the articulation of theory and the articulation of experiment. I shall arbitrarily call the more theoretical of these two activities 'calculation'. I do not mean mere computation, but the mathematical alteration of a given speculation, so that one brings it into closer resonance with the world.

Newton was a great speculator. He was also a great calculator; he invented the differential calculus in order to understand the mathematical structure of his speculation about the motions of the planets. Newton was also a gifted experimenter. Few scientists are great in all departments. P.S. Laplace (1749–1827) is an example of a supreme calculator. His celestial mechanics of about 1800 was, in its day, the sublime working out of Newton's theory of planetary motion. Newton had left innumerable questions unanswered, and it needed new mathematics to answer or even sometimes to ask the questions. Laplace put it all together in a remarkable way. He is also known as perhaps the greatest contributor to probability theory. At the beginning of a famous introductory lecture on probability, he states one classic version of determinism. He says that a supreme mind, given the equations of the universe, and a set of boundary conditions, would be able to work out the position and movement of every particle throughout all future time. One has the feeling that Laplace thought of this Supreme Being as a slightly superior version of Laplace, the Great Calculator. Laplace applied Newtonian ideas of attraction and repulsion to most topics, including heat and the velocity of sound. As I noted above just as Laplace was crowning Newton's achievement with mighty calculations, modest experimenters, with their Voltaic cells, compasses, and different coloured light filters were, to say the least, putting the Newtonian programme on hold.

The hypothetico-deductive scheme

My three-way distinction – speculation, calculation, and experiment – is not in conflict with traditional hypothetico-deductive accounts of science, such as N.R. Campbell's *Physics, the Elements* (1920, republished as *Foundations of Science*), as elaborated in R.B. Braithwaite's *Scientific Explanation* (1953). Campbell noticed that even in a finished theory, theoretical statements do not link directly with anything observable. There is no way to deduce experimental tests from, say, the central propositions of classical physics. Hence Campbell distinguished two layers of proposition. There are hypotheses, namely 'statements about some collection of ideas that are characteristic of the theory'. Then there is a 'dictionary' – Braithwaite calls it a Campbellian dictionary – of 'statements of the relation between these ideas and some ideas of a different nature'.

I disapprove of this distinction in linguistic terms of statement, but the idea has the ring of truth. It is closer to reality than the two-stage picture of conjecture and refutation. Campbell and Braithwaite indicate the answer to a puzzle. If speculation intends a qualitative structure for some domain, and experimentation, as I claim, sometimes pursues a life of its own, what then is the fit between the two? Answer: calculation *makes* the fairly tight hypothetico-deductive structure that you sometimes find in an elementary textbook. Calculators write the dictionary. They build the semantic bridge between theory and observation. Speculation and experiment need not in general be closely connected, but the activity I call calculation brings them close enough to discern a quantitative fit between the two.

I am not urging an exhaustive classification into three non-overlapping forms of life. I say only that the better version of the hypothetico-deductive story, with three tiers rather than two, is a hazy although not altogether hopeless snapshot of three kinds of ability that have to be distinguished within the mature and mathematized sciences.

Models

Reference to the hypothetico-deductive scheme shows that the division into speculation, calculation and experiment is conservative. The different levels of theoretical statement illustrated by the magneto-optical effect are not so unfamiliar. Nancy Cartwright's book, *How the Laws of Physics Lie* (1983), makes a more radical departure from tradition. Thus far I have written as if getting theory to mesh with possible determinations of nature is just a matter of articulation and calculation. We begin with speculations that we gradually cast into a form from whence experimental tests may be deduced. Not so. There is an enormously wide ranging intermediary activity best called model-building.

The word 'model' has come to mean different things in the sciences. In the early days of molecular biology, models of molecules were like scale models of aircraft that children make as a hobby. That is, they were bits of wire, wood, plastic and glue. I have seen attics full of discarded molecular biology models, made with spring washers, magnets, lots of tin foil and such. Some nineteenth-century physicists made similar hold-in-your-hand models of the inner constitution of nature, models built with pulleys, springs, string and sealing wax. Most generally, however, a model in physics is something you hold in your head rather than your hands. Even so, there is an odd mix of the pictorial and the mathematical. Take a look at a good text book, say N. Mott and I. Sneddon's *Wave Mechanics*. We find sentences like this:

The following idealized problem is instructive, although it does not refer to any actual physical phenomenon (p. 49).
We shall first treat the nucleus as of infinite mass (p. 54).
We treat the molecule as a rigid rod (p. 60).
We shall now calculate the energy levels of an electron in an atom when subjected to a magnetic field, without taking account of the spin (p. 87).
For *free* particles, however, we may take either the advanced or retarded potentials, or we may put the results in a symmetrical form, without affecting the result (p. 342).

The final quotation is grist to Cartwright's mill. Three models, at most one of which could (in logic) be true of the physical world, are used indifferently and interchangeably in a particular problem.

Roles for models

Suppose we say that there are theories, models, and phenomena. A natural idea would be that the models are doubly models. They are models of the phenomena, and they are models of the theory. That is, theories are always too complex for us to discern their consequences, so we simplify them in mathematically tractable models.

At the same time these models are approximate representations of the universe. In this picture, what Kuhn calls articulation becomes partly a matter of constructing models that human minds and known computational techniques can operate. That leads to the following conception.

1 The phenomena are real, we saw them happen.
2 The theories are true, or at any rate aim at the truth.
3 The models are intermediaries, siphoning off some aspects of real phenomena, and connecting them, by simplifying mathematical structures, to the theories that govern the phenomena.

In this picture, the phenomena are real and the theories aim at the truth, often being pretty close to the truth. Yes, there are examples of exactly that relationship. Cartwright remarks that there are also examples of many other kinds of relationship. She describes some in detail. Here I mention only two that she reports, without recapitulating her examples.

Realism about what?

The issues are closely connected with scientific realism. Cartwright is by and large anti-realist about theories. Models provide one basis for this. She notes that not only are models not deducible from the theory in which they are embedded, but that physicists can use, for purposes of convenience, a number of mutually inconsistent models within the same theory. Yet these models are the only available formal representations of the phenomenological laws that we think to be true. We have nothing more to go on, she says, than those phenomenological laws. Our formal modellings of them cannot be true altogether since they are not mutually consistent. Nor is there good reason to think one is all-round better than the other. None carry ground for belief back to the theory within which they are propounded. Moreover, models tend to be robust under theory change, that is, you keep the model and dump the theory. There is more local truth in the inconsistent models than in the more high-brow theories.

It may be said that this is a remark about the present stage of science. The realist, it is argued, speaks of a future, an ideal. We may converge to theories which by simplifying models we gradually connect with laws of phenomena. That is the truth at which we aim.

I respond to this in an inductive way. Every single year since 1840, physics alone has used successfully more (incompatible) models of phenomena in its day-to-day business, than it used in the preceding year. The ideal end of science is not unity but absolute plethora.

This remark can go along with intense admiration for projects which try to unify science. Faraday's discovery of the magneto-optical effect is a lesson for us all. Stephen Hawking, the great cosmologist, chose for the title of his 1980 inaugural lecture at Cambridge University, 'Is the end in sight for theoretical physics?' He thinks the answer is, yes. We shall have one unified theory. He added: that will leave most physics intact, for we shall still have to do applied physics, working out what happens from case to case.

Approximation

The relations of models to theory and to phenomena are various and complex. Approximations seem more straightforward. Cartwright shows that they are not. Our usual idea of an approximation is that we start with something true, and, to avoid mess, write down an equation that is only approximately true. But although there are such approximations *away* from the truth, there are far more approximations *towards* the truth. In many a theory of mathematical physics we have a structural representation with some equations at a purely hypothetical level, equations which are already simplifications of equations which cannot be solved. In order to make these fit some level of phenomenological law, there are endless possible approximations. After a good deal of fiddling someone sees that one approximation tallies nicely with the phenomena. Nothing in the theory says that this is the approximation we shall use. Nothing in the theory says that it is the truth. But it is the truth, if anything is. Cartwright alleges that the theory itself has no truth in it. It helps us think, but it is only representation. If there is any truth around, it lies in the approximations, not in the background theory.

The world

Cartwright concludes her introductory essay by referring to Pierre Duhem's 1906 distinction between two kinds of mind, the deep but narrow minds of the French and the broad but shallow minds of the English. (Leave aside the chauvinistic quibble that the deep mathematical physics of Duhem's day was done by Germans and the broad physical modelling referred to by Duhem was most often done by Scots. The Lagrange of this quotation was proud to be Italian.)

The French mind [she writes] sees things in an elegant, unified, way. It takes Newton's three laws of motion and the law of gravitation and turns them into the beautiful abstract mathematics of Lagrangian mathematics. The English mind, says Duhem, is an exact contrast. It engineers bits of gears, and pulleys, and keeps the strings from tangling up. It holds a thousand different details all at once, without imposing much abstract order or organization. The difference between the realist and me is almost theological. The realist thinks that the creator of the universe worked like a French mathematician. But I think that God has the untidy mind of the English (p. 19).

I myself prefer an Argentine fantasy. God did not write a Book of Nature of the sort that the old Europeans imagined. He wrote a Borgesian library, each book of which is as brief as possible, yet each book of which is inconsistent with every other. No book is redundant. For every book, there is some humanly accessible bit of Nature such that that book, and no other, makes possible the comprehension, prediction and influencing of what is going on. Far from being untidy, this is New World Leibnizianism. Leibniz said that God chose a world which maximized the variety of phenomena while choosing the simplest laws. Exactly so: but the best way to maximize phenomena and have simplest laws is to have the laws inconsistent with each other, each applying to this or that but none applying to all.

13 The creation of phenomena

One role of experiments is so neglected that we lack a name for it. I call it the creation of phenomena. Traditionally scientists are said to explain phenomena that they discover in nature. I say that often they create the phenomena which then become the centrepieces of theory.

The word 'phenomenon' has a long philosophical history. In the Renaissance some astronomers tried to 'save the phenomena', that is, produce a system of calculation that would fit known regularities. Not everyone admired that. Who will beat Francis Bacon's scorn when he writes, in a 1625 essay, *Superstition*: 'They are like astronomers, who did feign eccentrics and epicycles, and such engines of orbs, to save the phenomena; though they knew, there were no such things.' Yet the great French historian and philosopher of science, the eminent anti-realist Pierre Duhem, would admiringly take the same tag to name one of his books, *To Save the Phenomena* (1908). Bas van Fraassen recycles it for a chapter title in his book, *The Scientific Image*. Such authors teach that a theory provides a formalism to embed the phenomena within a coherent order, but the theory, where it extends beyond the phenomena, indicates no reality. They take for granted that the phenomena are discovered by the observer and the experimenter. How then can I say that a chief role for experiment is the creation of phenomena? Do I propound some sort of ultimate idealism in which *we* make the phenomena that even Duhem counts as 'given'? On the contrary, the creation of phenomena more strongly favours a hard-headed scientific realism.

Philological excursion

The word 'phenomenon' has an ancient philosophical lineage. In Greek it denotes a thing, event, or process that can be seen, and derives from the verb that means, 'to appear'. From the very beginning it has been used to express philosophical thoughts about appearance and reality. The word is, then, a philosopher's mine-field. Yet it has a fairly definite sense in the common writings of scientists. A phenomenon is *noteworthy*. A phenomenon is *discernible*. A phenomenon is commonly an event or process of a certain type that occurs regularly under definite circumstances. The word can also denote a unique event that we single out as particularly important. When we know the regularity exhibited in a phenomenon we express it in a law-like generalization. The very *fact* of such a regularity is sometimes called the phenomenon.

Despite this usage, many of the ancients held that phenomena are changing objects of the senses, as opposed to essences, the permanent reality. Thus phenomena were in contrast to reality. A present day positivist like van Fraassen holds that phenomena are the *only* reality. The word 'phenomenon' is neutral between these two doctrines.

Hellenistic writers contrasted phenomena to noumena, the things as they are in themselves. Kant transferred this to modern philosophy, and made the noumena unknowable. All natural science became a science of phenomena. Then came the cockcrow of positivism. The unknowable may be discounted, as if it did not exist. 'Phenomena' come to denote, for some empiricist philoso-phers, sense-data – private, personal, sensations. *Phenomenalism* is the doctrine that, in J.S. Mill's words, things are only the permanent possibilities of sensation, and that the external world is constructed out of actual and possible sense-data.

The word 'phenomenology' was introduced in 1764 by the physicist J.H. Lambert as the name for the science of phenomena, but the word has since split into two virtually distinct meanings. Philosophers will know that Hegel's *Phenomenology of the Spirit* (1807) is the study of how the mind develops through various stages of knowing itself as appearance but in the end grasps itself as reality. Early in this century 'phenomenology' was taken as the name for the German school of philosophy of whom Husserl is the most famous member. I was so trained in this philosophical sense of the word that when I lectured on the present topics in the Notre Dame *Perspectives* series (for which, much thanks) I was amazed to hear from the physics department there that they were hiring a phenomenologist. Phenomenology is an important part of solid state and particle physics. If you had wanted to check out what I wrote about muons and mesons in Chapter 8, you would likely have gone to some classic reference such as H. Bethe's *Mesons and Fields*. There you would have looked up muons and found the discussion followed by a long section on phenomenology. My use of the word 'phenomenon' is like that of the physicists. It must be kept as separate as possible from the philosophers' phenomenalism, pheno-menology and private, fleeting, sense-data. A phenomenon, for me, is something public, regular, possibly law-like, but perhaps exceptional.

Thus I pattern my use of the word after physics and astronomy. The Renaissance star-gazers meant both the regular observed motions of the spheres, and particular events, such as the occlusion of Mars, which they hoped would prove to be derived from some law-like structure of the heavens. But of course the astronomers were philosophers too, closer than we are to the Greek overtones of the word. Phenomena were 'appearances'. The historian of science Nicholas Jardine tells me that Kepler held it to be a defect of our solar system that when we look out, we behold phenomena – where the planets appear to move – rather than the true locations and paths of the heavenly bodies.

Solving the phenomena

Sometimes the old astronomers' talk of saving the phenomena was meant entirely seriously, but I think that often, long before Bacon, the usage was a little ironical. During the seventeenth century the scientific application of the word 'phenomenon' spread to any of what were called 'phenomena of nature'. This included both law-like regularities and what our modern insurance companies persist in calling acts of God: outstanding horrors like earthquakes. Daniel Defoe refers to the visibility of a star at noon as a phenomenon. A phenomenon could well be an anomaly rather than any known regularity.

The expression 'to save the phenomena' underwent some punning. It can be traced back to the Greek and then to Latin, where the word for 'save' would be *salve*. In the seventeenth century this got turned not into 'save' but into 'solve', so that David Hume, for example, will then write of 'the solution of the phenomenon'. That pretty well meant the *explanation* of the phenomenon, precisely the opposite of what Duhem means by saving the phenomena! Anyone who hopes that philology will teach some lesson to philosophy should feel chastened.

Has then the lineage of the word 'phenomenon' run so amok that there is no chance of attaching my sense to the word? On the contrary, the pedigree of my usage is surprisingly sound, as well as being the chief recent usage in natural science. During the eighteenth century, the English word 'phenomenon' was chiefly used my way. You might think that Berkeley would be a counter-example, for he is nowadays said to be a phenomenalist, reducing the external world to sense-data. Quite the contrary. Even toward the end of his career, when he wrote *Siris* (1744), we find the word over 40 times. This book is a marvellous although somewhat nutty tract about everything from constipation, through science, to belief in God. He uses the phrase, 'the phenomena of nature' in the standard manner of his time, to denote known regularities. It is true that Berkeley thought that all phenomena are appearances. But not because he thought they are sense-data! In the philosophical parts of the book Berkeley attempts to rebut the English natural philosophers who work in the traditions of Boyle and Newton. He provides a thoroughly immaterialist and somewhat anti-realist account of the solution of phenomena, but his remarks derive from his theories of matter and causation, not from some non-standard sense of the word in which 'phenomenon' itself denotes a sense-datum.

You cannot entirely rely on dictionaries here. That rich lode of examples, the *OED*, is often wrong on philosophical words because it reflects whatever anachronistic style of philosophy was then in fashion in the town where the great book was written. Thus the *OED* says that the word 'phenomenon' comes to mean 'the direct contents of sense experience' with the appearance of Thomas Reid's *Active Powers of the Human Mind* in 1788. That is a misreading of the very passage cited. Reid speaks of the phenomena of nature, and, like Berkeley, takes as his standard example the effect of a magnet on a compass. That effect is not some 'direct content of sense experience', as the dictionary puts it, but an observable regularity of nature. Reid is arguing the standard Newtonian line that becomes part of Comte's positivism: a solution of the phenomena provides descriptive laws but does not teach efficient causes.

It is to German philosophy that we owe the resurgence of the 'philosophical' sense of the word 'phenomenon', that is encoded both in the English school of phenomenalism and the continental school of phenomenology. Paradoxically had the British stuck with native masters such as Berkeley or Reid they would never have fallen into their own empiricist excesses.

Effects

When the physicists got their hands and minds on a truly instructive phenomenon, they came to call it an *effect*. I don't quite know when it began, but by the 1880s the practice became entrenched: the Faraday, or magneto-optical effect, the Compton effect, the Zeeman effect, the photoelectric effect, the anomalous Zeeman effect, the Josephson effect. Everitt notes that Maxwell speaks of the Peltier effect in his *Theory of Heat* (1872); perhaps that is where the usage began.

'Effects' were really starting to pile up in physics in the mid-1880s. One might use this as a symptom of a new stage in physics itself. What is an effect, and why do people come to call something an 'effect'? Let us take for example the effect discovered by E.H. Hall in 1879, while a research student in Rowland's new physics laboratory in Johns Hopkins University. Rowland had asked Hall to investigate a somewhat offhand remark of James Clerk Maxwell. In the *Treatise on Electricity and Magnetism*, Maxwell had said that when a conductor carrying a current is under the influence of a magnetic field, the field acts on the conductor but not on the current. In a recent study of the Hall effect, Jed Z. Buchwald uses the incident to recapture part of the spirit of Maxwellian theory of the time. Hall guessed that Maxwell was saying that the resistance of the conductor might be affected by the field, or that an electric potential might be produced. Hall failed to get the first effect, but finally located the second. He obtained a potential difference across a piece of gold leaf at right angles to both the magnetic field and the current. Some initial explanations of this turned out to be defective, because different conductors exhibit the effect of potential difference in a direction opposite to gold. Hall himself described the effect as a phenomenon – as does many a standard physics dictionary which, under the heading 'Hall effect', begins, 'the phenomenon which . . .' In his notebook entry for 10 November,

1879, Hall wrote, after describing some remarkable experimental successes, that:

> It seemed hardly safe, even then, to believe that a new *phenomenon* has been discovered, but now after nearly a fortnight has elapsed, and the experiment has been many times and under various circumstances successfully repeated . . . it is perhaps not too early to declare that the magnet does have an effect on the electric current or at least an effect on the circuit never before expressly observed or proved.[1]

Only a remark arising from within Clerk Maxwell's theoretical perspective would have put Hall on the lookout. What he found was not what Clerk Maxwell thought he might find. Nor was Hall testing a theory. This was exploration, as if Maxwell had said there might be some sort of island in those uncharted waters.

Phenomena and effects are in the same line of business: noteworthy discernible regularities. The words 'phenomena' and 'effects' can often serve as synonyms, yet they point in different directions. Phenomena remind us, in that semiconscious repository of language, of events that can be recorded by the gifted observer who does not intervene in the world but who watches the stars. Effects remind us of the great experiments after whom, in general, we name the effects: the men and women, the Compton and Curie, who intervened in the course of nature, to create a regularity which, at least at first, can be seen as regular (or anomalous) only against the further background of theory.

Creation

Hall did not create his effect! He discovered that passing a current through gold leaf, in a magnetic field, produces a potential at right angles to the field and to the current. He and other workers later investigated ramifications of the effect. What for example happens to conductors other than gold, or to semi-conductors? All this work required ingenuity. The apparatus was man-made. The inventions were created. But, we tend to feel, the phenomena revealed in the laboratory are part of God's handiwork, waiting to be discovered.

Such an attitude is natural from a theory-dominated philosophy. We formulate theories about the world. We conjecture various laws of nature. Phenomena are regularities, consequences of these laws.

1 Quoted by Jed. Z. Buchwald, *Centaurus* 23 (1979), p. 80.

Since our theories aim at what has always been true of the universe –
God wrote the laws in His Book, before the beginning – it follows
that the phenomena have always been there, waiting to be
discovered.

I suggest, in contrast, that the Hall effect does not exist outside of
certain kinds of apparatus. Its modern equivalent has become
technology, reliable and routinely produced. The effect, at least in a
pure state, can only be embodied by such devices.

That sounds paradoxical. Does not a current passing through a
conductor, at right angles to a magnetic field, produce a potential,
anywhere in nature? Yes and no. If anywhere in nature there is such
an arrangement, with no intervening causes, then the Hall effect
occurs. But nowhere outside the laboratory is there such a pure
arrangement. There are events in nature that are the resultant of the
Hall effect and of lots of other effects. But that mode of description –
the interaction or resultant of a number of different laws – is theory-
oriented. It says how we analyse complex events. We should not
have the picture of God putting in the Hall effect with his left hand
and another law with his right hand, and then determining the
result. In nature there is just complexity, which we are remarkably
able to analyse. We do so by distinguishing, in the mind, numerous
different laws. We also do so, by presenting, in the laboratory, pure,
isolated, phenomena.

We have the idea of numerous laws of nature adding up to a
'resultant'. That metaphor comes from mechanics. You have this
force and that force, this vector and that vector, and you can draw a
pretty diagram with ruler and compass to see what results. John
Stuart Mill remarked long ago that this fact about mechanics does
not generalize. Most science is not mechanics.

In the Renaissance the word 'phenomenon' denoted primarily
solar and astronomical regularities and anomalies. Long before God
had created the sun and the earth those who do not share my
Borgesian fantasy may imagine that He had some Universal Field
Theory in mind. When He made the heavens and the earth, they
obeyed gravitational and other field principles. The laws, we
imagine, had always been there. But the *phenomena* – or what the old
astronomers called the phenomena – did not exist until the creation
of our part of the universe. Likewise, I suggest, Hall's effect did not
exist until, with great ingenuity, he had discovered how to isolate,
purify it, create it in the laboratory. To update the example, 20 years

ago there were no masers or lasers in the universe. Perhaps that is wrong, perhaps there were one or two. (Some cosmological phenomena have recently been suggested as maser phenomena.) Yet now the universe has tens of thousands of lasers, many of them within three or four miles of me as I write.

The rarity of phenomena

It is no accident that in the Renaissance the word 'phenomenon' applied chiefly to celestial events. Nor is it by chance that nowadays the most respected ancient empirical science is astronomy. It is a good guess, although not a proven one, that a great variety of mammoth old earthworks, stone rings, Stonehenges, Mayan temples, scattered in all parts of the world, were built at enormous cost, to study the stars or the tides. Why did old science on every continent begin, it seems, with the stars? Because only the skies afford some phenomena on display, with many more that can be obtained by careful observation and collation. Only the planets, and more distant bodies, have the right combination of complex regularity against a background of chaos.

Did God provide no more phenomena for human beings to notice, than the heavens, the tides and other lunar phenomena such as menstruation? It will be protested that the world is full of manifest phenomena. All sorts of pastoral remarks will be recalled. Yet these are chiefly mentioned by city-dwelling philosophers who have never reaped corn nor milked a goat in their lives. (Many of my reflections of the world's lack of phenomena derive from the early morning milkstand conversations with our goat, Medea. Years of daily study have failed to reveal any true generalization about Medea, except maybe, 'She's ornery, often.') When I say that there are few phenomena in the world, the ample lore of mothers and hunters and sailors and cooks is cited in reply. Yet when we talk with romantics, who advise that we become wise and return to nature, we are not told to notice its phenomena but to become part of its rhythm. Moreover, most of the things called natural – yeast to make bread rise, for example – have a long history of technology.

Outside of the planets and stars and tides there are few enough phenomena in nature, waiting to be observed. Each species of plant and animal has its habits; I suppose each of those is a phenomenon. Perhaps natural history is as full of phenomena as the skies of night.

Every time I say that there are only so many phenomena out there in nature to be observed – 60, say – someone wisely reminds me that there are some more. But even those who construct the longest lists will agree that most of the phenomena of modern physics are manufactured. The phenomena about the species – say the one that a pride of lions hunts by having the male roar and sit at home base while the females chase after and kill scared gazelle – are anecdotes. But the phenomena of physics – the Faraday effect, the Hall effect, the Josephson effect – are the keys that unlock the universe. People made the keys – and perhaps the locks in which they turn.

The Josephson effect

It has long been known that about 4° above absolute zero, many funny things happen. Substances become superconductors, so that if using a heat switch you induce electricity in a closed circuit, the current keeps on going forever. What would happen if you separated superconductors by a thin sheet of electric insulation? What would happen if you had a battery connecting the two superconductors? Brian Josephson predicted in 1962 that a current flows between the two superconductors separated by an insulator. Moreover if you connect a battery there are wild oscillations of current with no net flow.

The Josephson effect is deduced from a theory of superconduction proposed five years earlier by J. Bardeen, J.N. Cooper and J.R. Schrieffer (the BCS theory). Superconduction is a movement of pairs of electrons, called Cooper pairs, which encounter no opposition in a cold body. For the current to stop, all the Cooper pairs must stop at the same time. That happens about as often as water boils in a refrigerator. When a supercold body warms up, the electrons separate and wander into an atom or whatever, and stop. Josephson realized that Cooper pairs would migrate across an insulator, constituting the Josephson current. Possibly this astonishing effect would not have been sought out had not BCS theory preceded it. Such a guess may be anachronistic (recent) history, for the basic idea is present in flux quantization, much discussed at the time. Only since then has flux quantization become an 'obvious' consequence of BCS theory. Whatever be the niceties of the facts, we notice something of a spectrum. Faraday found his magneto-optical effect because he hoped there ought to be some interaction between electromagnetism and light. Hall found his effect because Maxwellian electrodynamics suggested that one of two or three interactions ought to exist. Josephson found his effect by a brilliant deduction from the premises of theory. Hall did not 'confirm' Maxwellian theory, although he did add one more Maxwellian fact to the roster. Josephson really did confirm the new theory of superconduction. Note that this is not because the new theory provides the best explanation of the phenomenon. It is because no one would ever have thought of creating just that phenomenon without the theory.

I have switched language, in the last paragraph, from finding an effect to creating a phenomenon. That is deliberate. The Josephson effect did not exist in nature until people created the apparatus. The effect was not prior to theory. Talk about creating phenomena is perhaps made most powerful when the phenomenon precedes any articulated theory, but that is not necessary. Many phenomena are created after theory.

Experiments don't work

There is no more familiar dictum than that experimental results must be repeatable. On my view that works out as something of a tautology. Experiment is the creation of phenomena; phenomena must have discernible regularities – so an experiment that is not repeatable has failed to create a phenomenon.

Undergraduates and high school students know different. There is no more common comment on 'teaching evaluations' of courses with a laboratory component; the experiments do not work; the numbers have to be cooked, the reaction doesn't react, the phage does not grow. The laboratory just has to be improved!

Nor is this problem peculiar to the years of pre-apprenticeship. Here is another familiar story. My university has a very complex and expensive device X, of which there are few in the world; perhaps only ours works very well. It is the sort of device for which you book a year in advance, and are refereed by endless panels, before you are allowed to have two days working on X. Young hotshot A at our institution is obtaining some very striking results with X. Established figure B, in the same field, arrives for his two days and leaves frustrated. He even suggests we take a long hard look at A's work. Is A really getting what he claims to get? Or is he cheating? (This is a true story based on a tenure case that I reviewed.)

Now of course some laboratory courses are just awful. Sometimes old *B* has lost the knack or young *A* really is cheating. But as a paradoxical generalization one can say that most experiments don't work most of the time. To ignore this fact is to forget what experimentation is doing.

To experiment is to create, produce, refine and stabilize phenomena. If phenomena were plentiful in nature, summer blackberries there just for the picking, it would be remarkable if experiments didn't work. But phenomena are hard to produce in any stable way. That is why I spoke of creating and not merely discovering phenomena. That is a long hard task.

Or rather there are endless different tasks. There is designing an experiment that might work. There is learning how to make the experiment work. But perhaps the real knack is getting to know when the experiment is working. That is one reason why observation, in the philosophy-of-science usage of the term, plays a relatively small role in experimental science. Noting and reporting readings of dials – Oxford philosophy's picture of experiment – is nothing. Another kind of observation is what counts: the uncanny ability to pick out what is odd, wrong, instructive or distorted in the antics of one's equipment. The experimenter is not the 'observer' of traditional philosophy of science, but rather the alert and observant person. Only when one has got the equipment running right is one in a position to make and record observations. That is a picnic.

The pre-apprentice in the school laboratory is mostly acquiring or failing to acquire the ability to know when the experiment is working. All the thinking has been done, all the designing, all the implementation, but something is still missing. The ability to know when the experiment is working includes, of course, having sufficient sense of how this artifice works in order to know how to put it right. A laboratory course in which all the experiments worked would be fine technology but would teach nothing at all about experimentation. At the opposite end of the scale, it is not surprising that young hotshot *A* gets results and distinguished visitor *B* does not. *A* has had the opportunity to know the apparatus better; he has made part of it and suffered through its failures. That is an integral part of knowing how to create phenomena.

Repeating experiments

Folklore says that experiments must be repeatable. This has generated a philosophical pseudo-problem. It is clear that a variety of experiments is more compelling than repetitions of the same event. So philosophers have tried either to show that the repetitions are as valuable as the original, or have tried to explain, using say the calculus of probabilities, why the repetitions are less valuable. This is a pseudo-problem because, roughly speaking, no one ever repeats an experiment. Typically serious repetitions of an experiment are attempts to do the same thing better – to produce a more stable, less noisy version of the phenomenon. A repetition of an experiment usually uses different kinds of equipment. There are cases from time to time when people simply do not believe an experimental result and sceptics try again. Free quarks furnish such an example, as does the work on gravity waves. It was sensationally proposed 20 years ago that some wormy creatures could be got to do mazes; when others of their species ate the trained worms, the cannibals would do the mazes better too. This experiment was repeated because no one believed the result. Quite rightly, too.

In schools and colleges experiments are repeated *ad nauseam*. The point of those classroom exercises is never to test or elaborate the theory. The point is to teach people how to become experimenters – and to winnow out those for whom experimental science is not the right career.

It might seem as if there is one domain in which experiments must be repeated. That is when we are trying to make precise measurements, of, for example, constants of nature such as the velocity of light. We ought, it may seem, to make many determinations and average them out. How else could we determine that light travels at 299792.5 $\pm$ 0.4 kilometres per second? But even in this domain what is called for is a better experiment, not repetitions of less good trials on less good equipment. K.D. Froome and L. Essen write in their survey of *The Velocity of Light and Radio Waves* (p. 139):

We would repeat our philosophy of experimental measurement. The most important objective should be to increase the precision of measurement so that systematic errors can be measured and eliminated. Experience shows that extensive averaging processes invariably leave unsuspected systematic errors in the result. We see no advantage in taking a vast number of measurements as was done in the classical optical methods and in some of the recent determinations. We also regard it as unsound to take the standard deviation of the mean instead of that of a single observation as residual systematic errors are not reduced by taking more measurements. From the point of view of precision Froome's determination of 1958 is the only one to exceed those of Essen (1950) and Hansen and Bol (1950).

14 Measurement

We seem always to have measured. Were not Babylonian surveyors the precursors of geometry? Planetary observations, accurate to many places of sexagesimals, can be traced far back into the ancient world. Historians once said that Galileo was more a Platonist who did things in his head than an experimenter who did things with his hands, but they have since recovered some of his precise numerical observations of the acceleration of bodies on an inclined plane. We noticed Herschel spending a year of his mature life endlessly measuring reflections, refractions, degrees of transmissions of light or radiant heat. Hall's detection of the transverse electric potential required sensitive measurements of current. Measurements connected with Bragg's X-ray diffractions began the trip to molecular biology.

Since measurement is so obviously a part of scientific life, a little iconoclasm will do no harm. Did measurement always have its present role in physical science? Do we well understand the point of the most precise, delicate, and admired measurements in history? Is measurement an inherent part of the scientific mind, or does it stand for a philosophical position? Do measurements measure anything real in nature, or are they chiefly an artifact of the way in which we theorize?

Oddities

My most preposterous worry began when looking at a postcard in the Oxford History of Science Museum. It is a copy of a sixteenth-century painting called *The Measurers*. The curator must suppose it nicely complements his fine collection of brass instruments contemporary with the painting. A lady is measuring her cloth. A builder is measuring his gravel. An hour glass passes time. Sextants, astrolabes, and drafting instruments lie about. Yet nobody is measuring anything. The builders are paying no attention to the level of the gravel in their box. Sand drops in the hour glass unnoticed. The lady is holding her tape to the cloth, but not tautly. The tape loops down, so that the reading on the tape would be a foot longer than the length of the cloth.

Maybe this picture is a parody. Or perhaps the lady is only starting to measure the cloth. Someone is about to pick up the astrolabe. The builders are about to realize that the measuring box is running over. The hour glass will be noted soon. Or is it only we who, anachronistically, must read this painting in one of these two ways, as parody or as suspended beginning? Do we well understand old purposes of 'measuring'?

Herschel measured the proportions of light and heat transmitted by various substances to one part in a thousand. We doubt he could have had anything like that accuracy for light, and know it would have been impossible for heat. What was this cautious regular Newtonian inductivist doing, in 1800, with his wild exaggerations? His numbers were certainly not the result of applying a theory of errors. When we look to earlier times for the connection between numbers stated and observations made, historians are even more puzzled. Galileo may have been the first to think about averages, and it was a long time until the arithmetic mean – averaging – was a commonplace for experimenters. Gauss had provided a theory of error by 1807, and astronomers made use of that. Although all modern physical measurements demand an indication of error, physics outside of astronomy did not report estimates of error until the 1890s (or later).

Our conception of numbers and measuring is clear and unquestioned only at the end of the nineteenth century. After 1800 or so there is an avalanche of numbers, most notably in the social sciences. In his fundamental paper, 'A function for measurements in the physical sciences', Kuhn suggests that there was a second scientific revolution, during which a broad spectrum of physical science is, for the first time, 'mathematized'.[1] He puts this somewhere between 1800 and 1850. He suggests 1840 as a date when measuring, as we now conceive of it, takes on its fundamental role.

Constants of nature

Perhaps a turning point was signalled in 1832, the year that Charles Babbage (1792–1871), inventor of the digital computer, published a

---

[1] 'The function of measurement in modern physical science', in T.S. Kuhn, *The Essential Tension*, Chicago, 1979, pp. 178–224, esp. p. 220.

brief pamphlet urging publication of tables of all the constant numbers known in the sciences and arts. All the known constants would be printed. There are 20 categories of these. Babbage starts with a familiar list of astronomical quantities, specific gravities, atomic weights, and so forth. There are biological, geographical, and human numbers too: the lengths of rivers, the amount of oak a man can saw in an hour, the amount of air necessary to sustain human life for an hour, the mean length of bones of the several species, the number of students at universities and of books in great libraries.

Churchill Eisenhart, of the US Bureau of Standards, once suggested to me that Babbage's pamphlet marks the beginning of the modern idea of 'constants of nature'. He did not mean that constants had been unknown. Babbage himself lists many recent sources for this or that number. One fundamental constant, the G of Newtonian gravitation, had been known since at least 1798. The point is that Babbage sums up such work by stating officially, what was in the minds of many of his contemporaries, that the world might be defined by a set of numbers, which would be called constants.

Precise measurement

The everyday practice of measuring may need no explanation. Without measurement of a rather delicate sort, Hall could not have seen the effect of the current and field on the potential. He may have needed only a qualitative effect to start with, but without quite precise measurement, his successors could not have gone on to notice the differences among conductors, nor defined the 'Hall angle' as a characteristic of various substances. There is, however, another class of more memorable measurements that is problematic; it includes many of the great measurements of history.

We have to reconstruct texts to know much about Aristarchus's marvellous idea for finding the diameter of the earth by looking down a well at noon, and pacing across the desert. But we know a lot about how and why Cavendish 'weighed the earth' in 1798. Fizeau's 1847 work on the velocity of light is a masterpiece of precision. Its successor was the Michelson technique of diffraction gratings, which increased the potentiality for measurement by many orders of magnitude. Millikan's 1908–13 measurement of the charge on the electron is another milestone.

What is the point of these exceptional experiments? They are admired for at least two reasons. First, they were extraordinarily accurate. In no important way do we correct the figures of these pioneers. Secondly, each individual produced a brilliant new technique. Each experimenter had the genius not only to conceive a brilliant experimental idea, but also the gift to get it working, often by inventing numerous auxiliary experimental conceptions and technological innovations.

These two plain answers may not be quite good enough. What is the importance of the accuracy? What indeed is the point of this wonderful ingenuity in obtaining very accurate numbers that don't matter much? To begin with, let us not over-generalize. As always in the study of experiment, no one answer applies to every case.

The first consequence of Millikan's experiment is a qualitative confirmation that there is a minimum unit of electric charge. He found the charge on his oil droplets were small integral multiples of a single number. It was also inferred that this minimum charge must be the charge on the electron. Millikan expected as much, but in the days when electrons were in their infancy, that was a substantial result. The precise value of e, in that context, was as yet of little importance. In Millikan's own words, he had been able 'to present a direct and tangible demonstration that all electrical charges, however produced, are exact multiples of one definite elementary electrical charge . . .'. Millikan was of course also proud that he could 'make an exact determination of the value of the elementary electrical charge . . .'. Nor do I deny the words of the presentation speech for the Nobel Prize, that Millikan's 'exact evaluation of the unit has done physics an inestimable service, as it enables us to calculate with a higher degree of exactitude a large number of the most important physical constants'. However, if one is being iconoclastic about precise measurement, the power of a measurement to generate other measurements is hardly compelling justification.

One might well have doubted, in 1908, that there is a definite minimum negative charge e. But when Cavendish 'weighed the earth' in 1798, no one doubted that our planet does have a specific gravity. Cavendish's triumph was to measure this seemingly imponderable quantity. That not only satisfied intrinsic curiosity but also, by a short chain of inference, gave a value for the gravitational constant *G*. Newton had in fact known the answer all along (*Principia*, Book III, prop. x). He also suggested experiments, that were later done by a French expedition in Ecuador about 1740, which got pretty good results by noting the extent to which a plumb line is deflected from vertical when attracted by a big natural object such as the 6267 metre high Mt Chimborazo. Cavendish was more important because in determining *G* he was able to put into practice a new experimental idea (not original with him) in which one used artificial weights.

There is some analogy between the work done by Cavendish and Fizeau's 1847 measurements on the speed of light. In 1675 Roemer had estimated light's velocity from observations of the eclipses of the moons of Jupiter. His knowledge of planetary distance was poor, so he was wrong by 20% but (in analogy with Millikan) he had shown that there *is* a finite velocity of light that we now denominate *c*. At the end of the century Huygens had sufficient astronomy to give a good value for *c*. By 1847 the velocity of light was known, by Roemer's method, for any conceivable purpose.

What then was the point of Fizeau? It is of course important that different methods give the same results. Had Fizeau got an answer radically different from the Roemer method, we would have been plunged back into pre-Galilean astronomy, with light travelling at a different speed on earth than in the solar system. More importantly, Cavendish and Fizeau worked entirely in the laboratory with artificial instruments. You can't fool around with the moons of Jupiter, or Mt Chimborazo. This is connected with what I called the creation of phenomena. One is able to produce, in laboratory conditions, a stable numerical phenomenon over which one has remarkable control.

Fizeau did another experiment shortly after. How would the velocity of light be affected by passing through a tube of running water. Would the velocity simply be the sum of light and water speeds? His original point was connected with the theory of the aether, and some background is given in the next chapter. The last thing that Fizeau had in mind (or in 1852 could have had in mind) was a test between classical Newtonian theory and the theory of relativity. In his popular 1916 book, *The Theory of Relativity*, Einstein wrote of the two ways of summing motion, and continued: 'On this point we are enlightened by a most important experiment which the brilliant physicist Fizeau performed more than half a century ago, and which has been repeated since then by some of the best experimental physicists, so that there can be no doubt about its result'. Then Einstein remarks that a theory of this phenomenon was given by H.A. Lorentz, but continues, 'This circumstance does not in the least diminish the conclusiveness of the experiment as a crucial test of the relativity, for the electrodynamics of Maxwell–Lorentz, on which the original theory was based, in no way opposes the theory of relativity.' A remarkable statement: The experiment of over 50 years before was a crucial test for a brand new theory! The remark is doubly odd since traditional aether theory had no problem with Fizeau's result, and as we shall see in the next chapter, Michelson and Morley, who 'repeated' this sort of experiment in 1886, thought they had confirmed the existence of classical Newtonian aether. What we have is a brilliant mode of measurement which people put to their own ends. One end is whatever theory you like. Another is the development of even more ingenious variations on the technique, of which Michelson's work of 1881 has become the most famous instance. In this case we sometimes find the greatest theoretician, Einstein, glad for a moment to be a parasite, feeding haphazard on long dead experiments.

'Theory by other means'

Van Fraassen's *The Scientific Image* says that 'the real importance of theory, to the working scientist, is that it is a factor in experimental design' (p. 73). He proceeds to discuss Millikan and writes of this example that 'experimentation is the continuation of theory by other means'. These two remarks may seem at odds with each other. Perhaps he has a picture of experiment tugging along its own bootstraps, doing theory by other means in order that one can do more experimentation. That is not a bad picture of the Millikan example, because with a value for e, quite different experiments became possible.

The 'theory by other means' aphorism is based on the following idea. Theory had suggested that there is an electron and that electrons have a definite charge. But there is a blank in the theory; no theoretical reflection can fill in the value of e. We advance the theory 'by other means' by making an experimental determination of e.

This is an attractive metaphor, but I am reluctant to attach much weight to it. Cavendish filled in the value of the gravitational constant G, but he did not, I think, continue Newtonian theory one jot. Indeed we can look at it this way. Newtonian theory includes a statement about the gravitational force F existing between two masses m_1 and m_2 at a distance d from each other, namely:

$$F = G\frac{m_1 \, m_2}{d^2}$$

But the value of the constant G is simply not part of the theory. By filling in G Cavendish did not advance theory. As a matter of fact G is a unique constant of nature. As I shall shortly remark, most physical constants are connected by laws of physics to other constants. This is an important fact in determining each constant. G does not, however, relate to *anything* else at all.

Naturally we hope that G will turn out to relate to something. The gravitational force and the electromagnetic forces as well as the strong and weak forces may some day be embedded in a plausible theory. Or perhaps there is the following idea which pursues a 50-year-old speculation of P.A.M. Dirac. Suppose that the universe is about 10^{11} years old; then we might expect the gravitational force, compared to the electromagnetic force, to decrease by about 10^{-11} parts annually, a difference which is almost measurable with present technology. Such a measurement might teach us a lot about the world, but it would not be continuing Newtonian theory – or any other theory – by other means.

Millikan was more important to the theory of the electron, than Cavendish to the theory of gravity, but not because he filled in a blank in the theory. It was rather because he confirmed that there is a minimum unit of electron charge. It is evident by now that I share van Fraassen's repugnance to the model of science in which experimenters sit around waiting to be told to test, confirm, or refute theories. All the same, they often do confirm theories, even when, as in the case of Millikan, that is not the primary motivation. It seems to me that Millikan's relation to theory is that he confirmed a wide range of possible speculations to the effect that there is a minimum negative electric charge, likely associated with a con-jectural entity, the electron. He also found the value of that minimum charge, but that number does not have much to do with theory. Its payoff, as in the Nobel prize citation quoted above, lay in the fact that it helped fix other constants more precisely, but those constants too did not much influence the course of theory.

Are there exact constants of nature?

The only great philosopher familiar with measurement was C.S. Peirce, long employed by the US Coast and Geodesic Survey, and by the Lowell Observatory in Boston. He designed some nice pendulum experiments for determining G. Unlike the armchair philosopher he has nothing but scorn for the postulate that 'certain continuous quantities have exact values'. In 1892 he wrote in 'The doctrine of necessity reexamined' – an essay that occurs in most Peirce anthologies –

> To one who is behind the scenes, and knows that the most refined comparisons of masses, lengths, and angles far surpassing in precision all other measurements, yet fall behind the accuracy of bank accounts, and that the ordinary determinations of physical constants, such as appear from month to month in the journals, are about on a par with an upholsterer's measurements of carpets and curtains, the idea of mathematical exactitude being demonstrated in the laboratory will appear simply ridiculous (*The Philosophy of Peirce*, J. Buchler (ed.), (pp. 329f).

One finds a similar strand in Pierre Duhem. He regards the constants of nature as an artifact of our mathematics. We produce theories, which have various blanks in them, such as G. But it is not an objective fact about our universe that G is such and such. It is a qualitative fact that our universe can be represented by certain mathematical models, and from that there arises another qualitative fact, that there is something like an exact number that rides best with our mathematics. This is the basis of Duhem's pungent anti-realism about theories and natural constants.

Least squares adjustment

Do Duhem and Peirce betray a moment when constants are not exact? Not quite. Consider what has for the past decade been the most generally accepted set of fundamental constants, recommended for international use by the Committee on Data for Science and Technology.[2] The editors, Cohen and Taylor, have a very large

2 E.R. Cohen and B.N. Taylor, *Journal of Physical and Chemical Reference Data* 2 (1973), pp. 663–738.

number of fundamental constants, based on work at the main national laboratories throughout the world. Data are distinguished as 'More precise', 'Less precise WQED data' and 'Less precise QED data'. QED denotes work using theories of quantum electrodynamics, while WQED denotes work without that. Finally we have a few 'Other less precise quantities'. In the last section we find our friend the gravitational constant. The point about it is that 'at the present time, there exists no verified theoretical equation relating G to any other physical constant. Thus it can have no direct bearing on the output values of our adjustment' (p. 698).

What we mostly do with the other constants is to determine ratios between two constants. Thus the Josephson effect, discovered in 1962 (see Ch. 13) made a radical difference in precise measurement because it gave a curiously easy way to determine e/h, the relation between the charge on the electron and Planck's constant. By 1972 we knew, to five places of decimals, the precise value of the ratio of the mass of the electron and the muon; this in itself is determined from other ratios.

In the end we have a large number of numerical evaluations of constants. Then we proceed to that 'least squares fit'. We postulate that roughly speaking all the theories within a certain group are true (say QED or WQED). Thus we have a lot of equations connecting a lot of numbers. Naturally the numbers do not quite fit all the equations. Then we find an exact assignment of numbers, which makes all the equations true, and which minimizes the errors in all our best initial independent estimates of the various constants and ratios between constants. Naturally the business is a little more complex, for we attach different levels of accuracy in our initial measurements. This 'best fit', which comes with a built-in estimate of particular errors, then provides one assessment of all the constants, except a few loners such as the 'first' constant in science, namely G.

Factoring in the Josephson effect altered one set of previous estimates, which were all 'corrected'. The process is never ending:

However, since the publication of the 1973 adjustment, a number of new experiments have been completed, yielding improved values for some of the constants. . . . But it must be realized that, since the output values of a least-squares adjustment are related in a complex way and a change in the measured value of one of the constants usually leads to corresponding changes in the adjusted values of others, one must be cautious in carrying out calculations using both the output values from the 1973 adjustment and the results of more recent experiments.[3]

Undoubtedly when the next least squares adjustment is published (very soon), the whole web of theory and numbers will for a while seem more satisfactory. Yet a sceptic can insist that all we are doing is finding the most convenient set of numbers to plug into our constants. Perhaps our whole procedure can be cast into a Duhemian mould. At any rate we can hardly call this characteristic form of constant-determination 'continuing theory by other means'.

Measure everything

Kuhn says that the passion for measurement is fairly novel. He quotes Kelvin: ' I often say that when you can measure what you are speaking about, you know something about it; when you cannot measure it . . . your knowledge is of a meagre and unsatisfactory kind.'[4] Since Kelvin often said that, many garbled versions are in circulation. Karl Pearson recalls 'Lord Kelvin's statement that until you have measured a phenomena and turned it into numbers, you have a poor and vague apprehension of it.'[5] Should anyone think that the enthusiasm for measurement is untinged by ideology, consider this pastiche, in a long doggerel about the Ryerson lab in Chicago, which had become Michelson's base:

Now this is the law of Ryerson and this is the price of peace
That men shall learn to measure, or ever their strife shall cease.

Pearson, Kelvin, and the Ryerson Laboratory all come at the end of the nineteenth century. It began with an avalanche of numbers. The world was now conceived in a more quantitative way than ever before. The world is seen as constituted by numerical magnitudes. What were the effects of the fetish for measuring precise numbers on the course of natural science? To answer we should look at the essay by Kuhn already mentioned, 'A function for measurement in modern physical science', reprinted in his *The Essential Tension.*

---

3 From the pocket bible of high energy physics, *Particle Properties Data Booklet*, April 1982 (next edition April 1984), p. 3. Available from Lawrence Berkeley Laboratory and CERN.
4 William Thompson (Lord Kelvin), 'Electrical units of measurement', *Popular Lectures and Addresses*, London, 1889, Volume 1, p. 73.
5 K. Pearson, *The History of Statistics in the 17th and 18th Centuries*, London, p. 472.

The function of measurement

Why measure? One answer is Popper's dialectic of conjecture and refutation. Experiments, on that view, are intended to test theories. The best experiments put theories at greatest risk. Hence precise measurements must be the best experiments, because measured numbers are likely to conflict with predicted ones.

The child in Andersen's fairy tale said that the emperor wears no clothes. Kuhn is like that child. For all the finery of conjecture and refutation, the story imagined by Popper almost never happens. People don't make precise measurements in order to test theories. Cavendish did not test gravitational theory at all; he determined G. Fizeau got a better value for the velocity of light, and then used the technology that he had devised for that purpose to investigate (not to test) the possibility that light might have different speeds that would depend on the velocity of the medium in which it moved. Only 60 years later would Einstein suddenly find this to be a 'crucial test'. In more humdrum affairs, the numbers determined in the laboratory are not usually meant to put theory into jeopardy. Experiments, as Kuhn insists, are commonly rewarded when they get, with some precision, just the numbers people more or less expected them to get.

Most measurement, then, is what Kuhn called normal science. Good measurements demand new technology, and so invite much puzzle-solving of an experimental sort. Measurements articulate details of known material. Does it then follow that the fetishism of measurement that culminates in Kelvin had no effect on science except to intensify 'normal' activity? Not at all. Kuhn summarizes the function of measurement as follows: 'I believe that in the nineteenth century mathematization of physical science produced vastly refined professional criteria for problem solution and that it simultaneously very much increased the effectiveness of professional verification procedures' (p. 220). In a footnote he mentions 'the esoteric qualitative differences' that led to the selection of three problems: the photoelectric effect, black-body radiation, and specific heats. Quantum mechanics was the solution to these problems. Kuhn notes the speed with which the first version of quantum theory was accepted by the 'profession'. He gave us an unparalleled book on the second of these problems, that of *Black Body Theory and the Quantum Discontinuity 1894–1912*.

I gloss Kuhn as follows. We must distinguish the function of measurement from avowed reasons for measuring. Experimenters have various motives for measuring. They are rewarded when they devise ingenious systems of measurement. But the practice of measurement has a byproduct, by no means expected by Kelvin, Pearson, and the Ryerson Laboratory. Occasionally different batches of experimental numbers turn out not to mesh, contrary to expectation. That is an anomaly, sometimes even called an 'effect'. The greater the fetish for accuracy, the more often one will come across 'esoteric differences'. In fact not many do turn up, and these fascinating few anomalies provide the focus for professional problem-solving. When someone propounds a new theory its task is to explain the 'esoteric differences'. There are, then, quick tests which a new theory must meet. These are the effective verification procedures that Kuhn writes about, and they provide part of the structure of his view of scientific revolutions.

Let us not go overboard for this functional story. It is not the whole story. Of course many experiments are deliberately designed to test theories. Instrumentation is specifically developed in order to make the test more compelling. Nor is philosophy without effect. In Kelvin's day the old fact-finding positivism was rampant, and when one described one's experiment, one said one was trying to find hard numerical facts. Today, Popper's philosophy is rampant, and when one describes one's experiment, one says one is trying to test theories (otherwise you would not be funded!). Let us add too that Kuhn's account of measurement is not so different from Popper's. Precise measurement turns up phenomena that don't fit into theories and so new theories are proposed. But whereas Popper regards this as an explicit purpose of the experimenter, Kuhn holds it to be a byproduct. Indeed his account of this 'function' is very similar to what, in the social sciences, has been called functionalism.

Functionalism

Kuhn's philosophy is often said to turn into sociology. If that means empirical sociology, it is wrong. Kuhn has contributed no theorems like this: 'If a laboratory has more than N scientific personnel, the proportion of young scientists who enter the lab and are kept on to develop their careers is k; the proportion who pass on to other work is $1-k$.' Although Kuhn is no empirical sociologist, he is to some extent an old-fashioned speculative sociologist. Some of those, called functionalists, would discover a practice within a society or subculture. They would not ask how it got there, but why it stays. They conjectured that, given other aspects of the group, this practice has virtues that contribute to the preservation of the society itself. That is the function of the practice. It may well be unknown to the members of the society. But we should understand the practice in terms of its function.

Likewise, Kuhn notices that measurement plays an increasing role in physical science. Only by 1840, he suggests, do we find a thorough-going mathematization. He does not ask how that came to pass. He asks why it stayed. Cynics may suggest that measuring provides scientists with something to do. Kuhn says that the anomalies that inevitably turn up in a regimen of precise measurement focus subsequent activity, even in a state of what he calls crisis. They also determine what it is for a theory to be a good replacement to a previous one. Thus measurement has an important niche in Kuhn's view of normal science-crisis-revolution-new normal science.

An official view

Kuhn is curious and iconoclastic. Precise measurers of constants disregard his opinion, for the determination of constants seems to have become a world unto its own. Thanks to the Josephson effect, 'The U.S. National Bureau of Standards, on July 1, 1972 adopted the exact value $2e/h = 483593.420 \text{GHz/V}$ for use in maintaining the U.S. legal or as-maintained volt' (p. 667). There are at least 11 other as-maintained volts depending on the 11 major national laboratories in Japan, Canada etc. It is not quite mad to have 12 different regional 'volts', because part of the trouble is that when an experimenter wants to get a volt he has to go to the nearest lab, or employ 'shippable temperature-regulated volt transport standards'. Here is one philosophy of measurement: it comes at the end of the survey by Cohen and Taylor mentioned above, *The 1973 Least-Square Adjustment*: 'We believe that there is much useful work yet to be done in the fundamental constants field and that the romance of the next decimal place should be passionately pursued, not as an end in itself but for the new physics and the deeper understanding of nature that presently lie concealed there' (p. 726).

15 Baconian topics

Francis Bacon (1560–1626) was the first philosopher of experimental science.[1] Although he made no contribution to scientific knowledge, many of his methodological ideas are still with us. 'Crucial experiment' is an example.

He was a courtier born into the long reign of Elizabeth I. ('Being asked by the queen *how old he was*, he answered with much discretion, being then but a boy, *That he was two years younger than Her Majesty's happy reign.*')[2] He was the leading prosecuting attorney of his day, prosecuting 'criminal and capital alike'. ('He was never of an insulting and domineering nature over them, but always tender-hearted, . . . as one that looked upon the *example* with the eye of severity, but upon the *person* with the eye of pity and compassion.') He took bribes and was caught. ('I was the justest judge that was in England these 50 years: but it was the justest censure in Parliament these 200 years.')

He saw that observation of nature teaches less than experiment. ('The secrets of nature reveal themselves more readily under the vexation of art than when they go their own way.') He was something of a pragmatist. ('Truth therefore and utility are here the very same things, and works themselves are of greater value as pledges of truth than as contributing to the comforts of life.') He told us to experiment in order to 'shake out the folds of nature'. We must 'twist the lion's tail'. He quotes no sage more than Solomon: 'The glory of God is to conceal a thing; the glory of the king is to search it out.' He taught that in the true meaning of this proverb, every inquirer is king.

---

1 All quotations from Bacon in this chapter are from J. Robertson (ed.), *The Philosophical Works of Francis Bacon, reprinted from the texts and translations with the notes and prefaces of R.L. Ellis and F. Spedding*, London and New York, 1905. This is a selection from the standard *Works*.
2 These biographical snippets are from William Rawley's *Life of Bacon*, 1670, printed in the Bacon selection of the previous footnote.

Ant and bee

Bacon despised scholastic and bookish attempts to derive knowledge from first principles. We must instead create concepts and discover truths at a lower level of generality. Science should be built from the bottom up; Bacon did not foresee the value of the speculation, hypothesizing, and mathematical articulation, that we have since learned to use well in advance of any available system of testing. When he disdains writers who go beyond the facts, it is scholasticism, not the new science, that he has in mind. Hence he has been poorly treated by many among modern theory-dominated philosophers, who call him an inductivist. Yet it was Bacon who said that 'to conclude upon a bare enumeration of particulars (as the logicians do) without instance contradictory, is a vicious conclusion'. He called induction by simple enumeration puerile or childish.

Bacon, being a philosopher of experiment, does not fit well into the simple dichotomies of inductivism and deductivism. He sought to explore nature, for good or ill. 'No one should be disheartened or confounded if the experiments which he tries do not answer his expectation. For although a successful experiment be more agreeable, yet an unsuccessful one is oftentimes more instructive.' Thus Bacon already knew the value of learning by refutation. He sees that the new science will be an alliance of experimental and theoretical skills. In the manner of his time he draws a moral from insect life:

The men of experiment are like the ant; they only collect and use; the reasoners resemble spiders, who make cobwebs out of their own substance. But the bee takes a middle course; it gathers material from the flowers of the garden and the field, but transforms and digests it by a power of its own. Not unlike this is the true business of philosophy for it neither relies solely or chiefly on the powers of the mind, nor does it take the matter which it gathers from natural history and mechanical experiments and lay it up in the memory whole, as it finds it; but lays it up in the understanding altered and digested.

'Therefore,' he continues, 'from a closer and purer league between these two faculties, the experimental and the rational (such has never yet been made), much may be hoped.'

What's so great about science?

The alliance between the experimental and rational faculties had hardly begun when Bacon wrote so prophetically. In our day Paul Feyerabend asks, first, 'What is science?' and then, 'What's so great about science?' I do not find the second question all that pressing, but since we can sometimes see something rather grand in natural science, we can use Bacon to put our finger on it. Science is a league between those two faculties, the rational and the experimental. In Chapter 12 I divided Bacon's rational faculty into speculation and calculation, claiming that these are different abilities. What is so great about science is that it is a collaboration between different kinds of people: the speculators, the calculators, and the experimenters.

Bacon used to castigate the dogmatics and the empirics. The dogmatics were the men of pure theory. Many dogmatics of his day may have had the speculative cast of mind; some empirics must have been experimentalists of real talent. Each side alone produced little knowledge. What is characteristic of the scientific method? It brings these two abilities into contact by the use of a third human gift, the one I have called articulation and calculation. Even pure mathematics benefits from this collaboration. Mathematics was sterile after the Greek era, until it became 'applied' again. Even now, despite the power of much pure mathematics, many of the greatest contributors of deep 'pure' ideas – Lagrange, Hilbert, or whomever – were precisely those workers closest to the fundamental problems of the physical sciences of their time.

The remarkable fact about recent physical science is that it creates a new, collective, human artifact, by giving full range to three fundamental human interests, speculation, calculation, and experiment. By engaging in collaboration between the three, it enriches each in a way that would be impossible otherwise.

Hence we can diagnose doubts some of us share about the social sciences. Those fields are still in a world of dogmatics and empirics. There is no end of 'experimentation' but it as yet elicits almost no stable phenomena. There is plenty of speculation. There is even plenty of mathematical psychology or mathematical economics, pure sciences which have nothing much to do with either speculation or experimentation. Far be it from me to offer any evaluation of this state of affairs. Maybe all these people are creating a new kind of human activity. But many of us experience a sort of nostalgia, a feeling of sadness, when we survey social science. Perhaps this is because it lacks what is so great about fairly recent physical science. Social scientists don't lack experiment; they don't lack calculation; they don't lack speculation; they lack the collaboration of the three. Nor, I suspect, will they collaborate until they have real theoretical entities about which to speculate – not just postulated 'constructs' and 'concepts', but entities we can use, entities which are part of the deliberate creation of stable new phenomena.

Prerogative instances

Bacon's unfinished *Novum Organum* of 1620 has a curious classification of what he calls prerogative instances. These include striking and noteworthy observations. They include different kinds of measurement, and the use of microscopes and telescopes to extend our sight. They include the ways in which we reveal something intrinsically invisible, by means of its interaction with what we can observe. As I remarked in Chapter 10, Bacon does not speak of observation, nor does he think it important to distinguish instances which are simple seeings from those which are inferences drawn from delicate experiments. Indeed his use of instances is by and large more like the way in which modern physics speaks of observation, than the concept of observation found in positivist philosophy.

Crucial experiments

Bacon's fourteenth kind of instances are the *Instantiae crucis* – a term later rendered as crucial experiment. A more literal and perhaps more helpful translation would be, 'instances of the crossroads'. The old translators express it as 'instances of the fingerposts' for Bacon borrows 'the term from the fingerposts that are set up where roads part, to indicate the several directions'.

Later philosophy of science made crucial experiments absolutely decisive. The picture is that two theories are in competition, and then one single test conclusively favours one theory at the expense of the other. Even if the victorious theory is not proved true, at least the rival is knocked out of action. That is *not* what Bacon says about instances of the fingerposts. Bacon is truer than the more recent idea. He says that instances of the fingerposts 'afford very great light, and are of high authority, the course of interpretation *sometimes* ending in them and being completed'. I emphasize the word 'sometimes'. Bacon claimed only that crucial instances are sometimes decisive. It has recently become fashionable to say that experiments are crucial only with hindsight, that they never decide anything at the time. Imre Lakatos says just that. Hence a false confrontation has arisen. Had philosophers stuck with Bacon's good sense we might have avoided the following pair of contraries: (a) Crucial experiments decide decisively, and lead at once to the rejection of one theory; (b) 'There have been no crucial experiments in science' (Lakatos II, p. 211). Certainly Bacon disagrees with Lakatos, and rightly so, but he also dissents from (a).

Bacon's examples

Bacon's own examples are a mixed bag. Among instances of the fingerposts he includes some non-experimental data. Thus he considers a 'parting of the roads' concerning tides. Should we have the model of water shaking in a basin, now rising on one side, now on another? Or is it a lifting up of the water from the bottom, as when water, in boiling, rises and falls? So we ask the inhabitants of Panama whether the ocean ebbs and flows on opposite sides of the isthmus at the same time. What results is, as Bacon sees at once, not a decisive test, for there might be an auxiliary hypothesis saving one theory, one based on the rotation of the earth, for example. He then goes on to other considerations about the curvature of the oceans.

Bacon notes that most crucial instances are not furnished by nature: 'for the most part they are new, and are expressly and designedly sought for and applied, and discovered only by earnest and active diligence'. His nicest example concerns the problem of weight. 'Here the road will branch in two, thus: It must needs be that heavy and weighty bodies either tend of their own nature to the centre of the earth, by reason of their proper configuration; or else they are attracted by the mass and body of the earth itself.' Here is his experiment: take a pendulum clock, driven by lead weights, and a spring clock, synchronize them at ground level. Take them to a steeple or other high place, and later down a deep mine shaft. If the clocks do not keep equal time, it must be because of the effect of the weights and the distance of the attracting mass of the earth. It is a wonderful idea, although impracticable in Bacon's time. Presumably he would have obtained no effect, and thus favoured the false Aristotelian theory of proper motion. However, the fact that you were sent down the wrong road would not have upset Bacon too much. He never claimed that a crucial experiment must bring the task of interpretation to an end. You can always be sent down the wrong road and have to retrace your steps because the fingerposts are misleading.

Auxiliary hypotheses

Had Bacon's experiment been diligently tried in 1620, we must suppose that no one would have detected a difference between the pendulum and the spring clock. The instruments did not keep such good time anyway, and the deepest mine shaft and the tallest steeple in the same neighbourhood are not sufficiently far apart for the instruments to discriminate. A defender of gravitational theory could well reject the experimental result, asserting that finer measurements are needed.

That is the simplest way of saving an hypothesis from the negative result of a crucial experiment. It might seem that it is always possible to save an hypothesis in this way. Then there is the more general point made by the French philosopher and historian of science, Pierre Duhem. Whenever you test an hypothesis, you can at the same time save your preferred hypothesis by revising some auxiliary hypothesis connected with the method of testing. We have seen in Chapter 8 that Imre Lakatos thought this was a handy tool to sink the idea that hypotheses can be simply and directly falsified by experiment. As he puts it, 'exactly the most admired scientific theories simply fail to forbid any observable state of affairs' (I, p. 16). In support of this we get not fact but 'an imaginary case of planetary misbehaviour'. This makes the Duhemian point that one can commonly patch up a theory by adding auxiliary hypotheses; when one of the hypotheses pans out, that is a triumph for the theory, while if it does not, we just go on trying to get more auxiliaries. Thus, it is claimed, the theory does not forbid anything, for we get an inconsistency with observation only through intervening hypotheses. This too is ill-argued, and illustrates another kind of sloppiness. From the historical fact that hypotheses have sometimes been saved it is inferred that hypotheses can always be saved. This is argued not so much by an imaginary case as by an imaginative perversion of an historical event.

In 1814 and 1815 William Prout put forward two remarkable theses. At that time, following Dalton and others, precise determinations of atomic weights became possible. Prout proposed that all atomic weights are integral multiples of that of hydrogen, so that if we set $H = 1$, every other substance will have a whole number, as $C = 12$ or $O = 16$. The discrepancies between measurement and the whole numbers would then be experimental error. Secondly, all atoms would be made up of atoms of hydrogen. Thus hydrogen atoms would be the basic building blocks of the universe.

Prout was primarily a medical man with a taste for chemistry. He was one of several workers who, at about the same time, guessed at Avagadro's law. He discovered that there is HCl in the stomach, and that it plays a major role in digestion. He did some useful work on biological chemicals. He had no theoretical ground for his bold conjecture about hydrogen. Moreover it was *prima facie* false, for chlorine had an atomic weight of about 35.5. Lakatos uses Prout to illustrate how an hypothesis can stay afloat, wallowing in a sea of anomalies. He makes Prout into a significant figure who *knew* that chlorine had an atomic weight of 35.5, but still proposed that the weight is 'really' 36. He then 'corrects' this statement in a footnote. Actually Prout simply fudged the numbers to make it look as if they come out all right. But Lakatos is correct in saying that many able chemists in Britain stuck with Prout's hypothesis even when the figures looked bad. In continental Europe, where far more demanding chemical analysis was practised, far fewer people took Prout seriously.

Now we turn to auxiliary ways of saving an hypothesis. Lakatos says that you can never refute Prout, because you can just go on insisting that the chlorine has been imperfectly purified. So the real stuff has a weight of 36, although actual samples come out at 35.5. Lakatos gives us an imaginary statement, 'If seventeen chemical purifying procedures p_1, p_2, . . . p_{17} are applied to a gas, what remains will be pure chlorine.' Presented schematically we at once see that we can reject this, demanding that p_{18} be applied. But in real life it does not work like that. Worried that British (integral) atomic weights were at odds with continental ones, various committees were set up, and Edward Turner was commissioned to get to the heart of the matter. He regularly obtained 35.5, and for a while he was criticized, for example Prout suggested that the silver chloride might be carrying some water with it. A method was found to eliminate that possibility. It soon became clear to the community of British scientists that chlorine had an atomic weight of about 35.5. More sophisticated laboratories in Paris, still intrigued by the possibility that hydrogen is the building block of the universe, and shocked by having found that the old determinations for carbon are wrong, tried it all over again. But after much labour there was no possibility that chlorine had an atomic weight of 36. There was no way to save the hypothesis by hoping for better chemical purification, and that was that.

As it turned out, the hypothesis was on the verge of the truth, but that required a quite different research programme, and the idea of physical separation of the elements. At the beginning of our century, Rutherford and Soddy showed that elements do not have unique atomic weights, but are mixtures of different isotopes, so that the weight of 35.5 is an average of several real atomic weights. Moreover, Prout's second hypothesis is about right. If we speak not of hydrogen, but of the hydrogen ion, or proton, then the weights of all isotopes are essentially integral weights of that. It turns out not to be the only building block, but it is surely one of them.

We should not think of Prout's hypothesis as having been ' saved ' by auxiliary hypotheses. The process of eliminating analytical error simply came to an end. The atomic weight of chlorine on earth just is about 35.5 and nothing can alter that. As for the discovery of isotopes, that was not a new auxiliary hypothesis to save the Proutian so-called research programme. It was an entirely new hypothesis. Prout was merely the lucky chemist precursor of the physical idea. This has nothing to do with Duhem's thesis.

Crucial only with hindsight

Lakatos's opposition to crucial experiments denies the *un*-Baconian idea that there can be knockout tests favouring one theory and demolishing another. Only in retrospect, he says, do historians regard experiments as decisive. His methodology of research programmes teaches just that. If T is a current theory in programme

$P^\star$, we may devise an experiment to test T against $T^\star$. If T wins this round, it is still possible for $P^\star$ to recuperate and propose a better theory which in turn knocks out T. Only if, after some time, $P^\star$ gives up the ghost, do we later state that $T^\star$ was crucial.

In Bacon's more modest terminology, an experiment of the crossroads can be seen to be such at the time. If the trial favoured T, then the fingerposts say, truth may lie in the direction P. We can Lakatosize Bacon, much to the discomfort of both authors. Imagine a network of roads – an ordinary road map. At one intersection the fingerpost may say truth is in one direction, the direction of T and P. So we do not go down road $P^\star$. That road may still intersect with the P road later on. $P^\star$ puts up a revised theory $T_1^\star$. An instance of the fingerposts testing T and $T_1^\star$ may direct us now to follow the $P^\star$ road. Only if on the P road we never again intersect with $P^\star$ will we say, with hindsight, that the original crossroads was decisive.

This is, however, to play down the role of experiment too much. Certain types of experimental findings serve as benchmarks, permanent facts about phenomena which any future theory must accommodate, and which, in conjunction with comparable theoretical benchmarks, pretty permanently force us in one direction.

We can see this in the case of the controversial Michelson–Morley experiment. It was once cited as a decisive reason to reject the Newtonian idea, that space is filled with an all-pervading aether. Einstein replaced that by Einstein relativity. But he himself barely knew of the Michelson–Morley experiment, and its history is certainly not one of 'testing Newton and Einstein'. Lakatos uses this fact as a centrepiece in his onslaught on crucial experiments. He also uses it to argue that all experiment is subservient to theory.

In fact the experiment is a good example of the Baconian exploration of nature. It has been so much discussed that it will always be controversial, but it is useful to put an experimentalist version alongside that of Lakatos. To do that we must recall the aether from oblivion.

The all-pervading aether

Newton wrote 'All space is permeated by an elastic medium or aether, which is capable of propagating vibrations of sound, only with far greater velocity.' He went on to say that light is not a wave within the aether, but rather a medium through which light rays move. Newtonian optics made precious little use of the aether. Leibnizians were happy to mock it as an 'occult substance' just as they tried to dismiss gravity as an 'occult power'.

Waves: The wave theory really put aether to work. This was clearly stated by the founder (or re-inventor) of the wave theory, Thomas Young (1773–1829): '(I) a luminiferous Ether pervades the Universe, rare and elastic in a high degree. (II) Undulations are excited in this Ether whenever a body becomes luminous. (III) The sensation of different colours depends on the different frequency of Vibrations, excited by the Light in the Retina.'[3]

Aether wind: We owe the mathematics of the wave theory to Augustin Fresnel (1788–1827). He made the side assumption that if light were going through a medium, which itself was travelling in the opposite direction, then there would be a certain 'wind' effect – the apparent motion of the light would be diminished. This tallied in a vague way with the discovery in 1842 by J. Doppler (1803–53). If a light source is moving relative to the observer, then there is an alteration in the perceived frequency (colour) of the light. This is eminently a wave-like phenomenon, familiar from sound and the change in pitch associated in those days with train whistles and in ours with police sirens.

Astronomical aberration: Stars are not quite where they seem to be. This 'astronomical aberration' received several explanations. Fresnel obtained one from the aether-wind. In 1845 G.G. Stokes proposed the contrary idea that a moving body drags aether around with it. 'I shall suppose that the earth and the planets carry a portion of the ether along with them, so that the ether close to their surfaces is at rest relatively to their surfaces, while its velocity alters as we recede from the surface till, at no great distance, it is at rest in space.'[4]

Electromagnetism: James Clerk Maxwell brilliantly united the theory of light with that of electromagnetism. He was unenthusiastic about aether, but concluded that: 'Whatever difficulties we may

3 Thomas Young, 'Bakerian Lecture', *Philosophical Transactions of the Royal Society* 92 (1801), pp. 14–21.
4 G.G. Stokes, 'On the aberration of light', *Philosophical Magazine*, 3rd Ser., 27 (1845), pp. 9–10.

have in forming a consistent idea of the constitution of the ether, there can be no doubt that the interplanetary and interstellar spaces are not empty, but are occupied by a material substance or body . . .'[5] One of the problems was that no aether based on any variant of an elastic-solid model would work, that is, give the known laws of reflection and double refraction.

Wireless waves: In 1873 Maxwell predicted that there must be invisible electromagnetic waves, resembling light waves. H.R. Hertz (1857–94) vindicated Maxwell by eliciting radio waves. Hertz was somewhat dubious about aether, but in 1894 his great teacher, H. Helmholtz, would write posthumously of Hertz: 'By these investigations Hertz has enriched physics with new and most interesting views respecting natural phenomena. There can no longer be any doubt that light-waves consist of electric vibrations in the all pervading aether, and that the latter possesses the properties of an insulator and a magnetic medium.'[6]

Experiment

That is the briefest possible summary of the state of play around the time that Michelson commenced his now famous sequence of experiments. It is my purpose to contrast Lakatos's descriptions with those furnished by an experimenter. In 1878 Maxwell had written an article later to appear as 'Ether' in the ninth edition of the *Encyclopedia Britannica*. It suggests the idea for Michelson's experiment, at the same time implying that there is no hope of performing it.

If it were possible to determine the velocity of light by observing the time it takes to travel between one station and another on the earth's surface, we might, by comparing the observed velocities in opposite directions, determine the velocity of the ether with respect to these terrestrial stations. All methods, however, by which it is practicable to determine the velocity of light from terrestial experiments depend on the measurement of the time required by for the double journey from one station to the other and back again, and the increase of this time on account of the relative velocity of the

5 J. Clerk Maxwell, 'Ether', *Encyclopedia Britannica*, 9th edn., Volume 8 (1893), p. 572. (First circulated, 1878.)

6 H. von Helmholtz, Preface to H. Hertz, *The Principle of Mechanics* (D.E. Jones and J.J. Wallis, trans.), London, 1894, p. xi.

ether, equal to that of the earth in orbit would be only about one hundred millionth part of the whole time of transmission, and would therefore be quite insensible.[7]

Experimental idea: 'All methods', said Maxwell, would fail. Not so. Michelson realized that we should split a ray of light, by a half-silvered mirror, and send half the rays in the direction of the earth's motion, and the other at right angles to it. When they were reflected back we could see if there were any interference effect, because of a phase change caused by the two resultant light velocities. Hardly anyone believed that this would work. Michelson had difficulties too. For instance, horses going by outside completely upset the experiment by the otherwise unnoticeable jiggling of the building. In the end he went to the country and floated the whole experiment in a bath of mercury to damp out 'noise'. That is a characteristic experimental way of getting rid of unwanted phenomena.

Experiment to test theory: Lakatos writes: 'Michelson first devised an experiment in order to test Fresnel's and Stokes' contradictory theories about the influence of the motion of the earth on the aether.'

That is not true. As experimenter Michelson wanted to do what Maxwell said was impossible, namely measure the motion of the earth relative to the aether – regardless of anybody's theory. He says just that in a letter to Simon Newcomb, dated Berlin, 22 November, 1880. Michelson had studied in Paris under a pupil of Fizeau's and was ready for his own experimental determination. His patron was Alexander Graham Bell, to whom he wrote on 17 April, 1881: 'The experiments concerning the relative motion of the earth with respect to the ether have just been brought to a successful termination. The result was *negative*.'[8]

A negative result: The result was indeed negative. A positive result would have been sensational. For a positive result would have determined the absolute motion of the earth through space. If

7 Maxwell, 'Ether', p. 570.
8 Letter first published in Nathan Reingold, *Science in Nineteenth Century America*, Washington, 1971, pp. 288–90.

nature had only cooperated, this would have gone down in history as the triumph of centuries of speculation. We would know that space is absolute, and the absolute velocity with which the earth traverses space.

Result of the experiment: Lakatos writes: 'Michelson claimed that his 1881 experiment was a crucial experiment [between Fresnel and Stokes's explanations of aberration] and that it *proved* Stokes' theory.' Michelson said nothing of the sort. He wrote: 'The interpretation of these results is that there is no displacement of the interference bands. The result of the hypothesis of a stationary ether is thus shown to be incorrect, and the necessary conclusion follows that the hypothesis is erroneous.'[9] He did not claim to prove Stokes right, but at most that Fresnel was wrong.

Aberration: Michelson does continue by saying that his results 'directly contradict the explanation of the phenomenon of aberration generally accepted', that is, Fresnel's. At the end he says that 'it may not be out of place to add an extract' from a paper by Stokes. Stokes had said that there appears to be no 'result admitting of being compared with experiment, which would be different according to the theory we adopted' (i.e. Stokes's own, or Fresnel's). Stokes says, 'It would have been satisfactory if it had been possible to have put the two theories to the test of some decisive experiment.' Michelson quotes Stokes with deadpan no comment. He does *not* 'obliquely say' – as Lakatos puts it – that he had proved Stokes to be right. He does *not* call this a decisive experiment. What he implies is the experimenter's triumph over the theoretician: now I can determine that which has hitherto been inaccessible to you.

The 1886 experiment: Michelson teamed up with Morley, to redo Fizeau's experiment of 1852, in which light was sent through running water in a direction opposite to the flow of water. Morley came in as a chemist gifted in blowing glass, needed for the delicate glasswork for the running water. They concluded that Fizeau was chiefly right, although they somewhat reinterpreted Fresnel's theory. They ended by saying: 'The result of this work is therefore

9 A.A. Michelson, 'The relative motion of the earth and of the luminiferous ether', *American Journal of Science*, 3rd Ser., 22 (1881), p. 128.

that the result announced by Fizeau is essentially correct; and that the luminiferous ether is entirely unaffected by the motion of the matter which it permeates.'[10] I think Lakatos does not mention this experiment at all.

Theory enters: H. Lorentz, one of the great turn-of-the-century theoreticians, was keenly interested in aether. Lakatos somewhat overstates the case:

As often happens, Michelson the experimenter was then taught a lesson by a theoretician. Lorentz, the leading theoretical physicist . . . showed . . . that Michelson's calculations were wrong; Fresnel's theory predicted only half of the effect that Michelson calculated. . . . Indeed when a French physicist Potier, pointed out to Michelson his 1881 mistake, Michelson decided not to publish a correction note.

This is false. Michelson published the note in French in *Comptes Rendus* 94 (1882), p. 520. There was a footnote to Potier.

The 1887 experiment: This is the most famous Michelson–Morley experiment. Lakatos speaks of 'a letter from Rayleigh which draws attention to Lorentz's papers. This letter triggered off the 1887 experiment.' This is false. The letter was written early in 1887. The experiment was done in July 1887. You can see why Lakatos jumped to conclusions. In fact the experiment was planned for 1886 and fully funded then. Work was begun in October, but the literal foundations were destroyed in a fire of 27 October, 1886, thereby greatly delaying the execution. Thus the experiment was begun long before Rayleigh's so called triggering letter. (It may, however, have been triggered by Kelvin's lectures in Baltimore the year before.)

The 1887 experiment was in some ways less satisfactory than Michelson had hoped. With finer-honed equipment the two workers did not get a zero result. As Michelson wrote to Rayleigh in 1887, 'If the ether does slip past the earth, the relative velocity is less than one sixth of the earth's velocity.'[11] He thought they should redo the work at different times of year, and see if there were any discernible influence of altitude on aether draft. Lakatos finds it

10 A.A. Michelson and E.W. Morley, 'Influence of the motion of the medium on the velocity of light', *American Journal of Science*, 3rd Ser., 31 (1886).
11 Cf. R.S. Shankland, 'Michelson–Morley experiment', *American Journal of Physics* 32 (1964), pp. 16–35.

surprising that Michelson did not take up what he said one ought to do next. Was that because he was worried by what theory was doing? No. Michelson was an experimenter. He published a whole sequence of new work on his invention, the interferometer – work he found more fascinating than the aether. He captured the imagination of the American Association for the Advancement of Science with 'A plea for light waves' – waves which, using his invention, could provide a new way to define the standard meter.

Repeating the experiment: Michelson did twice return to the aether. Lakatos writes: 'Michelson's long series of experiments from 1881 to 1935, conducted in order to test subsequent versions of the ether program, provides a fascinating example of a degenerating problem shift.' Well, the experiments he did from 1931 to 1935 must be on the astral plane, for he died in 1931. The 'long series of experiments' between 1881 and 1935 done by Michelson are exactly these: 1881, 1886, 1887, 1897, 1925. Lots of other people tried to improve on or modify Michelson's results, but there is no long sequence of Michelson's experiments.

His 1897 experiment showed that altitude made no difference to his results, and he says there may be many explanations which he leaves the theoreticians to fuss about. Maybe, he says, the earth's atmosphere is bigger than we thought. Maybe the idea of the FitzGerald Contraction, just then in vogue, is correct. Maybe Stokes was right from the start. Michelson the experimenter is not pursuing any programme Lakatos writes of. As for the 1925 experiment, Miller claimed to detect an aether wind, so 75-year-old Michelson redid his youthful experiment to check out that he had not made a terrible mistake. He hadn't.

The experimental and rational faculties

Popper took the Michelson–Morley experiment as a clear crucial experiment bearing on the theory of relativity. In particular it invites the idea that light has the same velocity in all media and all directions. Lakatos and many others rightly say the historical relevance is tangential. Both Popper and Lakatos emphasize only the rational faculty. There are many more published fantasies about the Michelson–Morley experiment, and I certainly do not claim finality for my brief sketch. I chose Lakatos for an object lesson because I think his own philosophy is important. However, when it comes to drawing theoretical inferences from real-life cases, as with Prout or Michelson, the inference is always much too speedy. A theory-dominated philosophy blinds one to reality.

Doubtless Michelson is a little like Bacon's *ant*, a whizz at mechanical experiments and weak on theory – although not ignorant of it. Likewise, Lorentz was (to a lesser degree) a little like Bacon's *spider*. Both men thought highly of each other. Lorentz encouraged Michelson's work, while at the same time trying to develop an aether mathematics which would explain it away. If there was a degenerating programme it was, I suppose, Lorentz's. More importantly we see the interaction between two kinds of talents. The stupendous interest of Einstein's theories of relativity naturally makes the theoretical work the more important in this domain. Michelson too opened up new realms of experimental technique. Science, as Bacon wrote, must be like the *bee*, with the talents of both ant and spider, but able to do more, that is digest and interpret both experiments and speculation.

16 Experimentation and scientific realism

Experimental work provides the strongest evidence for scientific realism. This is not because we test hypotheses about entities. It is because entities that in principle cannot be 'observed' are regularly manipulated to produce a new phenomena and to investigate other aspects of nature. They are tools, instruments not for thinking but for doing. The philosopher's favourite theoretical entity is the electron. I shall illustrate how electrons have become experimental entities, or experimenter's entities. In the early stages of our discovery of an entity, we may test the hypothesis that it exists. Even that is not routine. When J.J. Thomson realized in 1897 that what he called 'corpuscles' were boiling off hot cathodes, almost the first thing he did was to measure the mass of these negatively charged particles. He made a crude estimate of e, the charge, and measured e/m. He got m about right, too. Millikan followed up some ideas already under discussion at Thomson's Cavendish Laboratory, and by 1908 had determined the charge of the electron, that is, the probable minimum unit of electric charge. Hence from the very beginning people were less testing the existence of electrons than interacting with them. The more we come to understand some of the causal powers of electrons, the more we can build devices that achieve well-understood effects in other parts of nature. By the time that we can use the electron to manipulate other parts of nature in a systematic way, the electron has ceased to be something hypothetical, something inferred. It has ceased to be theoretical and has become experimental.

Experimenters and entities

The vast majority of experimental physicists are realists about some theoretical entities, namely the ones they *use*. I claim that they cannot help being so. Many are also, no doubt, realists about theories too, but that is less central to their concerns.

Experimenters are often realists about the entities that they *investigate*, but they do not have to be so. Millikan probably had few qualms about the reality of electrons when he set out to measure their charge. But he could have been sceptical about what he would find until he found it. He could even have remained sceptical. Perhaps there is a least unit of electric charge, but there is no particle or object with exactly that unit of charge. Experimenting on an entity does not commit you to believing that it exists. Only *manipulating* an entity, in order to experiment on something else, need do that.

Moreover it is not even that you use electrons to experiment on something else that makes it impossible to doubt electrons. Understanding some causal properties of electrons, you guess how to build a very ingenious complex device that enables you to line up the electrons the way you want, in order to see what will happen to something else. Once you have the right experimental idea you know in advance roughly how to try to build the device, because you know that this is the way to get the electrons to behave in such and such a way. Electrons are no longer ways of organizing our thoughts or saving the phenomena that have been observed. They are ways of creating phenomena in some other domain of nature. Electrons are tools.

There is an important experimental contrast between realism about entities and realism about theories. Suppose we say that the latter is belief that science aims at true theories. Few experimenters will deny that. Only philosophers doubt it. Aiming at the truth is, however, something about the indefinite future. Aiming a beam of electrons is using present electrons. Aiming a finely tuned laser at a particular atom in order to knock off a certain electron to produce an ion is aiming at present electrons. There is in contrast no present set of theories that one has to believe in. If realism about theories is a doctrine about the aims of science, it is a doctrine laden with certain kinds of values. If realism about entities is a matter of aiming electrons next week, or aiming at other electrons the week after, it is a doctrine much more neutral between values. The way in which experimenters are scientific realists about entities is entirely different from ways in which they might be realists about theories.

This shows up when we turn from ideal theories to present ones. Various properties are confidently ascribed to electrons, but most of the confident properties are expressed in numerous different theories or models about which an experimenter can be rather agnostic. Even people in a team, who work on different parts of the same large experiment, may hold different and mutually incompatible accounts of electrons. That is because different parts of the experiment will make different uses of electrons. Models good for calculations on one aspect of electrons will be poor for others. Occasionally a team actually has to select a member with a quite different theoretical perspective simply in order to get someone who can solve those experimental problems. You may choose someone with a foreign training, and whose talk is well nigh incommensurable with yours, just to get people who can produce the effects you want.

But might there not be a common core of theory, the intersection of everybody in the group, which is the theory of the electron to which all the experimenters are realistically committed? I would say common lore, not common core. There are a lot of theories, models, approximations, pictures, formalisms, methods and so forth involving electrons, but there is no reason to suppose that the intersection of these is a theory at all. Nor is there any reason to think that there is such a thing as 'the most powerful non-trivial *theory* contained in the intersection of all the theories in which this or that member of a team has been trained to believe'. Even if there are a lot of shared beliefs, there is no reason to suppose they form anything worth calling a theory. Naturally teams tend to be formed from likeminded people at the same institute, so there is usually some real shared theoretical basis to their work. That is a sociological fact, not a foundation for scientific realism.

I recognize that many a scientific realism concerning theories is a doctrine not about the present but about what we might achieve, or possibly an ideal at which we aim. So to say that there is no present theory does not count against the optimistic aim. The point is that such scientific realism about theories has to adopt the Peircian principles of faith, hope and charity.[1] Scientific realism about entities needs no such virtues. It arises from what we can do at present. To understand this, we must look in some detail at what it is like to build a device that makes the electrons sit up and behave.

1 'I put forward three sentiments, namely interest in an indefinite community, recognition of the possibility of this interest being made supreme, and hope in the unlimited continuance of intellectual activity, as indispensable requirements of logic . . . these three sentiments seem to be pretty much the same as the famous trio of Charity, Faith and Hope . . .' C. Hartshorne and P. Weiss (eds.), *The Collected Papers of C.S. Peirce*, Volume 2, Section 665.

Making

Even if experimenters are realists about entities, it does not follow that they are right. Perhaps it is a matter of psychology: maybe the very skills that make for a great experimenter go with a certain cast of mind that objectifies whatever it thinks about. Yet this won't do. The experimenter cheerfully regards neutral bosons as merely hypothetical entities, while electrons are real. What is the difference?

There are an enormous number of ways in which to make instruments that rely on the causal properties of electrons in order to produce desired effects of unsurpassed precision. I shall illustrate this. The argument – it could be called the experimental argument for realism – is not that we infer the reality of electrons from our success. We do not make the instruments and then infer the reality of the electrons, as when we test an hypothesis, and then believe it because it passed the test. That gets the time-order wrong. By now we design apparatus relying on a modest number of home truths about electrons, in order to produce some other phenomenon that we wish to investigate.

That may sound as if we believe in the electrons because we predict how our apparatus will behave. That too is misleading. We have a number of general ideas about how to prepare polarized electrons, say. We spend a lot of time building prototypes that don't work. We get rid of innumerable bugs. Often we have to give up and try another approach. Debugging is not a matter of theoretically explaining or predicting what is going wrong. It is partly a matter of getting rid of 'noise' in the apparatus. Although it also has a precise meaning, 'noise' often means all the events that are not understood by any theory. The instrument must be able to isolate, physically, the properties of the entities that we wish to use, and damp down all the other effects that might get in our way. *We are completely convinced of the reality of electrons when we regularly set out to build – and often enough succeed in building – new kinds of device that use various well-understood causal properties of electrons to interfere in other more hypothetical parts of nature.*

It is not possible to grasp this without an example. Familiar historical examples have usually become encrusted by false theory-oriented philosophy or history. So I shall take something new. This is a polarizing electron gun whose acronym is PEGGY II. In 1978 it was used in a fundamental experiment that attracted attention even in *The New York Times*. In the next section I describe the point of making PEGGY II. So I have to tell some new physics. You can omit this and read only the engineering section that follows. Yet it must be of interest to know the rather easy-to-understand significance of the main experimental results, namely (1) parity is not conserved in scattering of polarized electrons from deuterium, and (2) more generally, parity is violated in weak neutral current interactions.[2]

Parity and weak neutral currents

There are four fundamental forces in nature, not necessarily distinct. Gravity and electromagnetism are familiar. Then there are the strong and weak forces, the fulfilment of Newton's programme, in the *Optics*, which taught that all nature would be understood by the interaction of particles with various forces that were effective in attraction or repulsion over various different distances (i.e. with different rates of extinction).

Strong forces are 100 times stronger than electromagnetism but act only for a minuscule distance, at most the diameter of a proton. Strong forces act on 'hadrons', which include protons, neutrons, and more recent particles, but not electrons or any other members of the class of particles called 'leptons'.

The weak forces are only 1/10 000 times as strong as electromagnetism, and act over a distance 1/100 times smaller than strong forces. But they act on both hadrons and leptons, including electrons. The most familiar example of a weak force may be radioactivity.

The theory that motivates such speculation is quantum electrodynamics. It is incredibly successful, yielding many predictions better than one part in a million, a miracle in experimental physics. It applies over distances ranging from the diameter of the earth to 1/100 the diameter of the proton. This theory supposes that all the forces are 'carried' by some sort of particle. Photons do the job in electromagnetism. We hypothesize 'gravitons' for gravity.

2 The popular account given below relies on generous conversations with some of the experimenters, and also on the in-house report, 'Parity violation in polarized electron scattering', by Bill Kirk, *SLAC Beam Line* no. 8 October, 1978.

In the case of interactions involving weak forces, there are charged currents. We postulate that particles called bosons carry these weak forces. For charged currents, the bosons may be positive or negative. In the 1970s there arose the possibility that there could be weak 'neutral' currents in which no charge is carried or exchanged. By sheer analogy with the vindicated parts of quantum electrodynamics, neutral bosons were postulated as the carriers in weak interactions.

The most famous discovery of recent high energy physics is the failure of the conservation of parity. Contrary to the expectations of many physicists and philosophers, including Kant, nature makes an absolute distinction between right-handedness and left-handedness. Apparently this happens only in weak interactions.

What we mean by right- or left-handed in nature has an element of convention. I remarked that electrons have spin. Imagine your right hand wrapped around a spinning particle with the fingers pointing in the direction of spin. Then your thumb is said to point in the direction of the spin vector. If such particles are travelling in a beam, consider the relation between the spin vector and the beam. If all the particles have their spin vector in the same direction as the beam, they have right-handed linear polarization, while, if the spin vector is opposite to the beam direction, they have left-handed linear polarization.

The original discovery of parity violation showed that one kind of product of a particle decay, a so-called muon neutrino, exists only in left-handed polarization and never in right-handed polarization.

Parity violations have been found for weak *charged* interactions. What about weak *neutral* currents? The remarkable Weinberg–Salam model for the four kinds of force was proposed independently by Stephen Weinberg in 1967 and A. Salam in 1968. It implies a minute violation of parity in weak neutral interactions. Given that the model is sheer speculation, its success has been amazing, even awe-inspiring. So it seemed worthwhile to try out the predicted failure of parity for weak neutral interactions. That would teach us more about those weak forces that act over so minute a distance.

The prediction is: Slightly more left-handed polarized electrons hitting certain targets will scatter, than right-handed electrons. Slightly more! The difference in relative frequency of the two kinds of scattering is one part in 10 000, comparable to a difference in probability between 0.50005 and 0.49995. Suppose one used the standard equipment available at the Stanford Linear Accelerator in the early 1970s, generating 120 pulses per second, each pulse providing one electron event. Then you would have to run the entire SLAC beam for 27 years in order to detect so small a difference in relative frequency. Considering that one uses the same beam for lots of experiments simultaneously, by letting different experiments use different pulses, and considering that no equipment remains stable for even a month, let alone 27 years, such an experiment is impossible. You need enormously more electrons coming off in each pulse. We need between 1000 and 10 000 more electrons per pulse than was once possible. The first attempt used an instrument now called PEGGY I. It had, in essence, a high-class version of J.J. Thomson's hot cathode. Some lithium was heated and electrons were boiled off. PEGGY II uses quite different principles.

PEGGY II

The basic idea began when C.Y. Prescott noticed (by 'chance'!) an article in an optics magazine about a crystalline substance called gallium arsenide. GaAs has a curious property. When it is struck by circularly polarized light of the right frequencies, it emits lots of linearly polarized electrons. There is a good rough and ready quantum understanding of why this happens, and why half the emitted electrons will be polarized, 3/4 polarized in one direction and 1/4 polarized in the other.

PEGGY II uses this fact, plus the fact that GaAs emits lots of electrons due to features of its crystal structure. Then comes some engineering. It takes work to liberate an electron from a surface. We know that painting a surface with the right stuff helps. In this case, a thin layer of cesium and oxygen is applied to the crystal. Moreover the less air pressure around the crystal, the more electrons will escape for a given amount of work. So the bombardment takes place in a good vacuum at the temperature of liquid nitrogen.

We need the right source of light. A laser with bursts of red light (7100 Ångstroms) is trained on the crystal. The light first goes through an ordinary polarizer, a very old-fashioned prism of calcite, or Iceland spar. This gives linearly polarized light. We want circularly polarized light to hit the crystal. The polarized laser beam now goes through a cunning device called a Pockel's cell. It electrically turns linearly polarized photons into circularly polarized ones. Being electric, it acts as a very fast switch. The direction of circular polarization depends on the direction of current in the cell. Hence the direction of polarization can be varied randomly. This is important, for we are trying to detect a minute asymmetry between right- and left-handed polarization. Randomizing helps us guard against any systematic 'drift' in the equipment. The randomization is generated by a radioactive decay device, and a computer records the direction of polarization for each pulse.

A circularly polarized pulse hits the GaAs crystal, resulting in a pulse of linearly polarized electrons. A beam of such pulses is manoeuvred by magnets into the accelerator for the next bit of the experiment. It passes through a device that checks on a proportion of polarization along the way. The remainder of the experiment requires other devices and detectors of comparable ingenuity, but let us stop at PEGGY II.

Bugs

Short descriptions make it all sound too easy, so let us pause to reflect on debugging. Many of the bugs are never understood. They are eliminated by trial and error. Let us illustrate three different kinds: (1) the essential technical limitations that in the end have to be factored into the analysis of error; (2) simpler mechanical defects you never think of until they are forced on you; (3) hunches about what might go wrong.

1 Laser beams are not as constant as science fiction teaches, and there is always an irremediable amount of 'jitter' in the beam over any stretch of time.

2 At a more humdrum level the electrons from the GaAs crystal are back-scattered and go back along the same channel as the laser beam used to hit the crystal. Most of them are then deflected magnetically. But some get reflected from the laser apparatus and get back into the system. So you have to eliminate these new ambient electrons. This is done by crude mechanical means, making them focus just off the crystal and so wander away.

3 Good experimenters guard against the absurd. Suppose that dust particles on an experimental surface lie down flat when a

polarized pulse hits them, and then stand on their heads when hit by a pulse polarized in the opposite direction? Might that have a systematic effect, given that we are detecting a minute asymmetry? One of the team thought of this in the middle of the night, and came down next morning frantically using antidust spray. They kept that up for a month, just in case.

Results

Some 10^{11} events were needed to obtain a result that could be recognized above systematic and statistical error. Although the idea of systematic error presents interesting conceptual problems, it seems to be unknown to philosophers. There were systematic uncertainties in the detection of right- and left-handed polarization, there was some jitter, and there were other problems about the parameters of the two kinds of beam. These errors were analysed and linearly added to the statistical error. To a student of statistical inference this is real seat-of-the-pants analysis with no rationale whatsoever. Be that as it may, thanks to PEGGY II the number of events was big enough to give a result that convinced the entire physics community. Left-handed polarized electrons were scattered from deuterium slightly more frequently than right-handed electrons. This was the first convincing example of parity violation in a weak neutral current interaction.

Comment

The making of PEGGY II was fairly non-theoretical. Nobody worked out in advance the polarizing properties of GaAs – that was found by a chance encounter with an unrelated experimental investigation. Although elementary quantum theory of crystals explains the polarization effect, it does not explain the properties of the actual crystal used. No one has got a real crystal to polarize more than 37% of the electrons, although in principle 50% should be polarized.

Likewise although we have a general picture of why layers of cesium and oxygen will 'produce negative electron affinity', that is, make it easier for electrons to escape, we have no quantitative understanding of why this increases efficiency to a score of 37%.

Nor was there any guarantee that the bits and pieces would fit together. To give an even more current illustration, future experimental work, briefly described below, makes us want even more electrons per pulse than PEGGY II could give. When the parity experiment was reported in *The New York Times*, a group at Bell Laboratories read the newspaper and saw what was going on. They had been constructing a crystal lattice for totally unrelated purposes. It uses layers of GaAs and a related aluminium compound. The structure of this lattice leads one to expect that virtually all the electrons emitted would be polarized. So we might be able to double the efficiency of PEGGY II. But at present that nice idea has problems. The new lattice should also be coated in work-reducing paint. The cesium–oxygen compound is applied at high temperature. Hence the aluminium tends to ooze into the neighbouring layer of GaAs, and the pretty artificial lattice becomes a bit uneven, limiting its fine polarized-electron-emitting properties. So perhaps this will never work. Prescott is simultaneously reviving a souped-up new thermionic cathode to try to get more electrons. 'Theory' would not have told us that PEGGY II would beat out thermionic PEGGY I. Nor can it tell if some thermionic PEGGY III will beat out PEGGY II.

Note also that the Bell people did not need to know a lot of weak neutral current theory to send along their sample lattice. They just read *The New York Times*.

Moral

Once upon a time it made good sense to doubt that there are electrons. Even after Thomson had measured the mass of his corpuscles, and Millikan their charge, doubt could have made sense. We needed to be sure that Millikan was measuring the same entity as Thomson. More theoretical elaboration was needed. The idea needed to be fed into many other phenomena. Solid state physics, the atom, superconductivity: all had to play their part.

Once upon a time the best reason for thinking that there are electrons might have been success in explanation. We have seen in Chapter 12 how Lorentz explained the Faraday effect with his electron theory. I have said that ability to explain carries little warrant of truth. Even from the time of J.J. Thomson it was the measurements that weighed in, more than the explanations. Explanations did help. Some people might have had to believe in electrons because the postulation of their existence could explain a wide variety of phenomena. Luckily we no longer have to pretend to infer from explanatory success (i.e. from what makes our minds feel good). Prescott *et al.* don't explain phenomena with electrons. They know how to use them. Nobody in their right mind thinks that electrons 'really' are just little spinning orbs around which you could, with a small enough hand, wrap the fingers and find the direction of spin along the thumb. There is instead a family of causal properties in terms of which gifted experimenters describe and deploy electrons in order to investigate something else, for example weak neutral currents and neutral bosons. We know an enormous amount about the behaviour of electrons. It is equally important to know what does not matter to electrons. Thus we know that bending a polarized electron beam in magnetic coils does not affect polarization in any significant way. We have hunches, too strong to ignore although too trivial to test independently: for example dust might dance under changes of direction of polarization. Those hunches are based on a hard-won sense of the kinds of things electrons are. (It does not matter at all to this hunch whether electrons are clouds or waves or particles.)

When hypothetical entities become real

Note the complete contrast between electrons and neutral bosons. I am told that nobody can yet manipulate a bunch of neutral bosons, if there are any. Even weak neutral currents are only just emerging from the mists of hypothesis. By 1980 a sufficient range of convincing experiments had made them the object of investigation. When might they lose their hypothetical status and become commonplace reality like electrons? When we use them to investigate something else.

I mentioned the desire to make a better gun than PEGGY II. Why? Because we now 'know' that parity is violated in weak neutral interactions. Perhaps by an even more grotesque statistical analysis than that involved in the parity experiment, we can isolate just the weak interactions. That is, we have a lot of interactions, including say electromagnetic ones. We can censor these in various ways, but we can also statistically pick out a class of weak interactions as precisely those where parity is not conserved. This would possibly give us a road to quite deep investigations of matter and anti-matter. To do the statistics one needs even more electrons per pulse than

PEGGY II could hope to generate. If such a project were to succeed, we should be beginning to use weak neutral currents as a manipulable tool for looking at something else. The next step towards a realism about such currents would have been made.

Changing times

Although realisms and anti-realisms are part of the philosophy of science well back into Greek prehistory, our present versions mostly descend from debates about atomism at the end of the nineteenth century. Anti-realism about atoms was partly a matter of physics: the energeticists thought energy was at the bottom of everything, not tiny bits of matter. It also was connected with the positivism of Comte, Mach, Pearson and even J.S. Mill. Mill's younger associate Alexander Bain states the point in a characteristic way in his textbook, *Logic, Deductive and Inductive*. It was all right for him to write in 1870 that:

Some hypotheses consist of assumptions as to the minute structure and operation of bodies. From the nature of the case these assumptions can never be proved by direct means. Their merit is their suitability to express phenomena. They are Representative Fictions.

'All assertions as to the ultimate structure of the particles of matter,' continues Bain, 'are and ever must be hypothetical. . . .' The kinetic theory of heat, he says, 'serves an important intellectual function'. But we cannot hold it to be a true description of the world. It is a Representative Fiction.

Bain was surely right a century ago. Assumptions about the minute structure of matter could not be proved then. The only proof could be indirect, namely that hypotheses seemed to provide some explanation and helped make good predictions. Such inferences need never produce conviction in the philosopher inclined to instrumentalism or some other brand of idealism.

Indeed the situation is quite similar to seventeenth-century epistemology. At the time knowledge was thought of as correct representation. But then one could never get outside the representations to be sure that they corresponded to the world. Every test of a representation is just another representation. 'Nothing is so much like an idea as an idea,' as Bishop Berkeley had it. To attempt to argue for scientific realism at the level of theory, testing, explanation, predictive success, convergence of theories, and so forth is to be locked into a world of representations. No wonder that scientific anti-realism is so permanently in the race. It is a variant on the 'spectator theory of knowledge'.

Scientists, as opposed to philosophers, did in general become realists about atoms by 1910. Despite the changing climate, some anti-realist variety of instrumentalism or fictionalism remained a strong philosophical alternative in 1910 and in 1930. That is what the history of philosophy teaches us. The lesson is: think about practice, not theory. Anti-realism about atoms was very sensible when Bain wrote a century ago. Anti-realism about *any* sub-microscopic entities was a sound doctrine in those days. Things are different now. The 'direct' proof of electrons and the like is our ability to manipulate them using well-understood low-level causal properties. I do not of course claim that reality is constituted by human manipulability. Millikan's ability to determine the charge of the electron did something of great importance for the idea of electrons: more, I think, than the Lorentz theory of the electron. Determining the charge of something makes one believe in it far more than postulating it to explain something else. Millikan gets the charge on the electron: better still. Uhlenbeck and Goudsmit in 1925 assign angular momentum to electrons, brilliantly solving a lot of problems. Electrons have spin, ever after. The clincher is when we can put a spin on the electrons, polarize them and get them thereby to scatter in slightly different proportions.

There are surely innumerable entities and processes that humans will never know about. Perhaps there are many that in principle we can never know about. Reality is bigger than us. The best kinds of evidence for the reality of a postulated or inferred entity is that we can begin to measure it or otherwise understand its causal powers. The best evidence, in turn, that we have this kind of understanding is that we can set out, from scratch, to build machines that will work fairly reliably, taking advantage of this or that causal nexus. Hence, engineering, not theorizing, is the best proof of scientific realism about entities. My attack on scientific anti-realism is analogous to Marx's onslaught on the idealism of his day. Both say that the point is not to understand the world but to change it. Perhaps there are some entities which in theory we can know about only through theory (black holes). Then our evidence is like that furnished by

Lorentz. Perhaps there are entities which we shall only measure and never use. The experimental argument for realism does not say that only experimenter's objects exist.

I must now confess a certain scepticism, about, say, black holes. I suspect there might be another representation of the universe, equally consistent with phenomena, in which black holes are precluded. I inherit from Leibniz a certain distaste for occult powers. Recall how he inveighed against Newtonian gravity as occult. It took two centuries to show he was right. Newton's aether was also excellently occult. It taught us lots. Maxwell did his electromagnetic waves in aether and Hertz confirmed the aether by demonstrating the existence of radio waves. Michelson figured out a way to interact with the aether. He thought his experiment confirmed Stokes's aether drag theory, but in the end it was one of many things that made aether give up the ghost. The sceptic like myself has a slender induction. Long-lived theoretical entities, which don't end up being manipulated, commonly turn out to have been wonderful mistakes.[3]

3 On p. 272 above, weak neutral bosons are used as an example of purely hypothetical entities. In January 1983 CERN announced observing the first such particle W in proton–antiproton decay at 540 GeV.

Further reading

There is an annotated bibliography of 95 items at the end of my anthology of some post-Kuhnian philosophy of science:
(1)   Ian Hacking (ed.), *Scientific Revolutions*, Oxford, 1982.
I shall not duplicate that here, nor list books already prominently discussed above. For the chapters of Part A 'Representing', here are a few classics, some useful anthologies, and some recent writing. A few of the anthologies are numbered in order to make it easy to refer back to them. Since few of the topics in Part B 'Intervening' have been much discussed by philosophers, I do not attempt a chapter by chapter breakdown, but direct attention to a few essays that I found helpful.

Introduction: Rationality

The place to start is, of course,
T.S. Kuhn, *The Structure of Scientific Revolutions*, Chicago, 1962, 2nd edn, with postscript, 1969.
Kuhn's essays on related topics are found in,
> *The Essential Tension: Selected Studies in Scientific Thought and Change*, Chicago, 1977.
> 'Commensurability, comparability, communicability', *PSA 1982*, Volume 2.
> 'What are scientific revolutions?' Occasional Paper no. 18, Center for Cognitive Science, Massachusetts Institute of Technology.
An excellent anthology of essays about Kuhn's ideas is,
(2)   Gary Gutting (ed.), *Paradigms and Paradoxes*, Notre Dame, 1980.
Here are three books and a collection of essays about rationality in science.
Larry Laudan, *Progress and its Problems*, California, 1977.
W. Newton-Smith, *The Rationality of Science*, London, 1981.
Husain Sarkar, *A Theory of Method*, California, 1983.
(3)   Martin Hollis and Steven Lukes (eds.), *Rationality and Relativism*, Oxford, 1982.
One should also consult the work associated with Imre Lakatos, listed for Chapter 8 below. A thorough study of the history of the idea of scientific revolution is:
I.B. Cohen, *Revolution in Science: The History, Analysis and Significance of a Concept and a Name*, Cambridge, Mass., 1984.

1 What is scientific realism?

For an excellent overview of the current debate, see,
(4)   Jarrett Leplin (ed.), *Essays on Scientific Realism*, Notre Dame, 1983.

There are now a great many classifications of scientific realisms. One is:
Paul Horwich, 'Three forms of realism', *Synthese* 52 (1982), pp. 181–201.

2 Building and causing

In addition to *Sense and Sensibilia*, cited in the text, one may find other examples of Austin's treatment of English words in:
J.L. Austin, *Philosophical Papers*, 3rd edn, Oxford, 1979.
Despite the initial influence of this work, I regret to report that almost nobody does that kind of philosophy today. Austin also had a more speculative programme, which has been adapted by some influential philosophers in Germany, and, to a lesser extent, in the United States:
How to do Things with Words, Oxford, 1963.
For harsh criticism of what Austin says about the word 'real', read:
Jonathan Bennett, 'Real', in K. Fann (ed.), *J.L. Austin, A Symposium*, London, 1969.
Smart's own introductory textbook is:
J.J.C. Smart, *Between Science and Philosophy: An Introduction to the Philosophy of Science*, New York, 1968.
It is not clear that Cartwright's causalism has exact forbears, but she acknowledges substantial debts to the anti-realist classic, originally published in French in 1906,
Pierre Duhem, *The Aim and Structure of Physical Theory*, Princeton, 1954.
In an at present unpublished discussion note that I have just seen, Bas van Fraassen claims that causalism has its roots in Newton's search for *vera causa* (true causes) combined with the famous assertion, *hypotheses non fingo* (I do not make, or depend upon, hypotheses).

3 Positivism

As noted in the text, many trace the positivist spirit back to Hume or earlier. Still, the word is Comte's. Any University library will have in its catalogue several books of selections from Comte in translation. One of the figures most often cited as a positivist is Ernst Mach. That is not a closed case. Paul Feyerabend will contribute a long essay to a Grover Maxwell memorial volume (University of Minnesota Press, expected 1984) in which he contends vigorously that Mach was no positivist. A reading of Mach would well begin with,
Ernst Mach, *The Analysis of Sensations*, Chicago, 1887, and numerous reprintings, with several variations on the title.
A more clearcut classic of positivism is,
Karl Pearson, *The Grammar of Science*, London, in numerous and substantially altered or augmented editions, from 1897 on.
The classic criticism of positivism in that stage of its evolution singles out Pearson as the one positivist whose empirical good sense makes him stop short of the excesses of his peers:

V.I. Lenin, *Materialism and Empirio-Criticism*, New York, 1923.
The best anthology of logical positivism is:
A.J. Ayer (ed.), *Logical Positivism*, New York, 1959.

4 Pragmatism

The most interesting historical survey of pragmatism is,
Bruce Kuklick, *The Rise of American Philosophy: Cambridge, Massachusetts, 1860–1930*, New Haven, 1977.
There are numerous anthologies of Peirce, James, and Dewey. A new and more satisfactory edition of Peirce's writing is well in hand and at least two computer concordances of his surviving,works are increasingly available. Any established anthology will, however, provide a pretty good account of his philosophy for all but the specialist scholar. His essays are in my opinion so popular and yet so deep that they improve with rereading every couple of years or so.

5 Incommensurability

The debate about incommensurability was due to discussions by Paul Feyerabend as well as Kuhn:
Paul Feyerabend, 'On the meaning of scientific terms', *The Journal of Philosophy* 62 (1965), pp. 266–74.
   'Problems of empiricism', in R. Colodny (ed.), *Beyond the Edge of Certainty*, Englewood Cliffs, N.J., 1965.
   Against Method, London, 1977.
   Science in a Free Society, London, 1979.
Among the very many discussions of incommensurability, one may especially note:
Dudley Shapere, 'The structure of scientific revolutions', *The Philosophical Review* 73 (1964), pp. 383–94. Reprinted in (2).
   'Meaning and scientific change', in R. Colodny (ed.), *Mind and Cosmos: Essays in Contemporary Science and Philosophy*, Pittsburgh, 1966, pp. 41–85. Reprinted in (1).
Hartrey Field, 'Theory change and the indeterminacy of reference', *The Journal of Philosophy* 70 (1973), pp. 462–81.
G. Pearce and P. Maynard (eds.), *Conceptual Change*, Dordrecht, 1973.
Arthur Fine, 'How to compare theories: reference and change', *Noûs* 9 (1975), pp. 17–32.
Michael Levine, 'On theory-change and meaning-change', *Philosophy of Science*, 46 (1979).

6 Reference and 7 Internal realism

Many of the papers in (4) contain useful studies of or allusions to Putnam, whose views about realism have notoriously evolved in the course of time.

It is important to read his collected papers in chronological order; likewise with his books.

Hilary Putnam, *Mind, Language and Reality; Philosophical Papers*, Volume 2, Cambridge, 1979.

Meaning and the Moral Sciences, London, 1978.

History, Truth and Reason, Cambridge, 1981.

Views which in some ways overlap Putnam's have long been urged by Nelson Goodman, who summarizes them in,

Nelson Goodman, *Ways of Worldmaking*, Indianapolis, 1978.

Putnam's more formal presentation of the Löwenheim–Skolem argument about realism is given in,

'Models and reality', *The Journal of Symbolic Logic* 45 (1980), pp. 464–82.

Numerous discussions of this argument will be appearing soon.

G.R. Merrill, 'The model-theoretic argument against realism', *Philosophy of Science* 47 (1980), pp. 69–81.

J.L. Koethe, 'The stability of reference over time', *Noûs* 16 (1982), pp. 243–52.

M. Devitt, 'Putnam on realism, a criticial study of Hilary Putnam's *Meaning and the Moral Sciences*', *Noûs*, forthcoming.

David Lewis, 'New work for a theory of universals', *The Australasian Journal of Philosophy*, forthcoming.

8 A surrogate for truth

Many of Lakatos's views about science are foreshadowed in a highly original and entertaining dialogue on the nature of mathematics.

Imre Lakatos, *Proofs and Refutations: The Logic of Mathematical Discovery*, Cambridge, 1976.

In 1965 he organized a conference involving Popper, Carnap, Kuhn and numerous others. The third and most lively volume of this conference contains his own most important contribution to the philosophy of science.

I. Lakatos and A. Musgrave (eds.), *Criticism and the Growth of Knowledge*, Cambridge, 1970.

Two memorial volumes discussing the work of Lakatos and its applications are:

Colin Howson (ed.), *Method and Appraisal in the Physical Sciences*, Cambridge, 1976.

R. S. Cohen *et al.* (eds.), *Essays in Memory of Imre Lakatos*, Dordrecht, 1976.

Break: Reals and representations

Since no bibliography fits the subject matter of the break I shall take the opportunity of drawing attention to two interesting schools that deploy social studies of science to draw philosophical conclusions. In Edinburgh we find the very strong doctrine that nearly all scientific reality is a social construct. The paper, 'Relativism, rationalism and the sociology of knowledge', contained in (3) above, provides a rich list of sources. Some of the main statements by this group are:

Barry Barnes, *Scientific Knowledge and Sociological Theory*, London, 1974. *Interests and the Growth of Knowledge*, London, 1977.

David Bloor, *Knowledge and Social Imagery*, London, 1976.

Some support for this group is found in the second chapter of a very innovative set of essays,

Mary Hesse, *Revolutions and Reconstructions in the Philosophy of Science*, Brighton, 1980.

At Bath there is another group of sociologically oriented students of science, who have valuable things to contribute to the second half of the present book, 'Intervening', for they have made internal studies of a variety of experimental work, ranging from parapsychology to laser physics.

H.M. Collins and T.J. Pinch, *Frames of Meaning: The Social Construction of Extraordinary Science*, London, 1981.

H.M. Collins, 'The TEA set: tacit knowledge and scientific networks', *Science Studies* 4 (1974), pp. 165–86.

H.M. Collins and T.G. Harrison, 'Building a TEA laser: the caprices of communication', *Social Studies of Science* 5 (1975), pp. 441–50.

David Gooding, 'A convergence of opinion on the divergence of lines: Faraday and Thomson's discussion of diamagnetism', *Notes and Records of the Royal Society of London* 36 (1982), pp. 243–59.

H.M. Collins, 'Son of seven sexes: the social destruction of a physical phenomenon', *Social Studies of Science* 11 (1981), pp. 33–62.

The last paper describes the rejection of some experimental results in the investigation of gravity waves.

9–16 Intervening

For an analysis of Millikan's work on the electron, see,

G. Holton, *The Scientific Imagination*, Cambridge, 1978, Chapter 2.

Holton urges that Millikan's use of data is strongly influenced by theoretical expectations. For a summary of this and related aspects of Holton's work, see,

'Thematic presuppositions and the direction of scientific advance', in A.F. Heath (ed.), *Scientific Explanation*, Oxford, 1981, pp. 1–27.

This volume also contains a strong statement of the position of the theoretician, by A. Salam (cf. p. 267, above): 'The nature of the "ultimate" explanation in physics', *ibid.*, pp. 28–35. Here is a case history of a crucial experiment, together with a detailed account of that experiment, and a philosophical discussion of 'good' experiments:

Allan Franklin and Howard Smokler, 'Justification of a "crucial" experiment: parity nonconservation', *American Journal of Physics* 49 (1981), pp. 109–11.

Allan Franklin, 'The discovery and nondiscovery of parity nonconservation', *Studies in History and Philosophy of Science* 10 (1979), pp. 201–57.

'What makes a good experiment?' *British Journal for the Philosophy of Science* 32 (1981), pp. 367–74.

There are few books studying experimental histories in detail. One of the best is about the discovery of isotopes by E. Rutherford and F. Soddy. The same author has two interesting papers about two different ways that a science can, for a while, go off in the wrong direction.

Thaddeus Trenn, *The Self-Splitting Atom*, London, 1975.

'Thoruranium (U-236) as the extinct natural parent of thorium: the premature falsification of an essentially correct theory', *Annals of Science* 35 (1978), pp. 581–97.

'The phenomenon of aggregate recoil: the premature acceptance of an essentially incorrect theory', *Annals of Science* 37 (1980), pp. 81–100.

A blow-by-blow account of the Michelson–Morley experiment is given by:

Loyd S. Swenson, *The Etherial Aether: A History of the Michelson–Morley Experiment*, Austin, Tex., 1972.

On causes, models and approximations, see,

R. Harré, *Causal Powers: A Theory of Natural Necessity*, Oxford, 1975.

M. Hesse, *Models and Analogies in Science*, London, 1963.

In addition to the works by Hesse cited on pp. 162 and 280 above, two other books by these authors will be found useful,

R. Harré, *The Philosophers of Science: An Introductory Survey*, Oxford, 1972.

M. Hesse, *Forces and Fields: The Concept of Action at a Distance in the History of Physics*, Westport, Conn., 1970.

The most recent contributor to the history and philosophy of new experimental physics is publishing the following papers. The first has a bearing on my account of muons and mesons (pp. 87–90, above), and the second on weak neutral currents (Chapter 16, above):

Peter Galison, 'The discovery of the muon and the failed revolution against quantum electrodynamics', *Centaurus*, April, 1983.

'How the first neutral current experiments ended', *Reviews of Modern Physics*, April, 1983.

'Einstein's experiment, the g-factor, and theoretical predispositions', *Historical Studies in the Physical Sciences* 12 (1982), pp. 285–323.

Index